Fire Under Control

FIRE UNDER CONTROL

CORE PRINCIPLES OF STRUCTURAL FIREFIGHTING

STEVE BERNOCCO

DISCLAIMER

The recommendations, advice, descriptions, and methods in this book are presented solely for educational purposes. The author and publisher assume no liability whatsoever for any loss or damage that results from the use of any of the material in this book. Use of the material in this book is solely at the risk of the user.

Fire Engineering Books & Videos
110 S. Hartford Ave., Suite 200
Tulsa, Oklahoma 74120 USA

800.752.9764
+1.918.831.9421
info@fireengineeringbooks.com
www.FireEngineeringBooks.com

Senior Vice President: Eric Schlett
Operations Manager: Holly Fournier
Sales Manager: Joshua Neal
Managing Editor: Mark Haugh
Production Manager: Tony Quinn
Developmental Editor: Chris Barton
Cover Designer: Brandon Ash
Book Designer: Robert Kern, TIPS Technical Publishing, Inc.
Compositor: Alyssa Taylor, TIPS Technical Publishing, Inc.
Cover Photo: Mark Blair

Library of Congress Control Number: 2020944224

ISBN print 978-1-593704-94-0
ISBN epub 978-1-593709-62-4

Printed in the United States of America

1 2 3 4 5 25 24 23 22 21

This book is dedicated to all firefighters everywhere, who put their lives on the line every day for people they may or may not know. I am so very proud to be associated with such a noble group of individuals.

And to the men and women of the Seattle Fire Department—thanks for all of it!

Capt. B's 5 Keys to Success in the Fire Service

1. Know your job
2. Train every shift
3. Stay fit
4. Be kind
5. Teamwork gets it done

CONTENTS

FOREWORD

As a firefighter, nothing gets the blood flowing, the mind engaged, and the muscles firing quite like a good fire. I know, I know; I'm in the fire service and I am not supposed to talk like that, but here I am speaking to fire service folks, my people, and you all know it's true. Fighting fire is what we do. It's what we want to do. It's what we live to do. Getting on the end of a hoseline, forcing a door, cutting a roof, searching for those citizens we are sworn to protect, man, that stuff is incredible. It is the difficult but fulfilling work that Steve Bernocco has brought to life in this book.

In my opinion, hitting a fire hard and fast is a personal "self-actualization" moment . . . the top of Maslow's Hierarchy of Needs . . . the pinnacle . . . a firefighter's moment of Zen.

That's what this book, *Fire Under Control* is about—the lessons and the stories take the reader to the tip of the spear in firefighting, to the smoke and the heat. Steve Bernocco has constructed a great framework for developing, or improving, your own fireground skill set.

The mission is to save lives. Every firefighter takes an oath to that mission. To meet the mission, every firefighter must approach the work like Steve. You must have a willingness to seek out knowledge on a continuing basis. *Fire Under Control* brings that knowledge and skill to life. Foundational principles for effective firefighting include knowledge of building construction, fire behavior, hydraulics, and ventilation. A commitment to training. Willingness to work as part of a team. *Fire Under Control* brings those concepts to life for the firefighter, the company officer, and the chief officer.

To apply knowledge, you must have the skill to match it. This is where training comes to bear. Steve and I met on the training ground over two decades ago. As lieutenants in the Seattle Fire Department Recruit Training Program, we were tasked with taking new civilian hires and turning them into firefighters in twelve weeks. I recognized immediately his passion for training, individual improvement, and team chemistry. I have been learning from him ever since. *Fire Under Control* gives you access to the methods, or perhaps the madness, that Steve developed during decades as a committed firefighter, fire officer, and trainer. This book provides you the opportunity to stand on the shoulder of a giant and look into your future in the fire service. It takes the reader from knowledge, to training, to implementation on the fireground. The path is clear and convincing. Steve has the chops to bring it home: decades of training, decades of service, excellence in writing. Learn his tools, and his applications, to improve your own skill and your team's effectiveness, and meet the mission of your fire department more effectively.

Enjoy this book. I know I did.

—Phillip Jose
Deputy Chief (ret.), Seattle Fire Department

Acknowledgments

Many people have contributed their thoughts, expertise, and time to help me make this book become a reality. And though I cannot thank everyone involved, I would like to recognize those who have been instrumental.

First, I would like to thank the development editor of Fire Engineering Books, Chris Barton, who believed in me and supported my vision of how this book would look and read. His guidance and help were invaluable throughout the process. I also want to thank Mark Haugh, who oversees Fire Engineering Books and Videos. Mark heard my initial idea and gave it a green light, giving Chris and I the permission to try something different and fresh. Thanks, Mark! And lastly, I must thank Tony Quinn, the production manager for Fire Engineering Books and Videos, and his entire team. Tony believed in me, gave me expert guidance and feedback, and allowed me to try something completely new. Tony and his team at were instrumental in taking my raw manuscript and shaping it into what it is today. Thanks!

Next, I would like to thank John Odegard, whose phenomenal fireground photos grace the cover and pages of this book. John understands structural firefighting—his father was a firefighter in the Tacoma (WA) Fire Department. John always captures truly amazing action shots of Seattle firefighters at work, and he is dedicated to his fireground photography and to the Seattle Fire Department. This book would not be as engaging without John's excellent photos, and I thank him for his allowing me to use them and for his friendship.

Steve Kerber, from UL and the College Park Fire Department, helped me early on with the Fire Behavior section of my manuscript. Thanks, Steve! Along with Dan Madrzykowski from NIST, your dedication and hard work have changed the way we understand how fire behaves in structures.

My good friend Deputy Chief (ret.) Phil Jose helped me by reading and editing my early manuscript, offering his thoughts and ideas, which were invaluable. Phil has been my most honest critic ever since we first came up with the idea of bringing air management to the fire service many years ago. He has pushed me to become a better writer and instructor, and I thank him for his insight, guidance, and friendship. Thanks for writing the foreword and showing me that "It's About Them!"

Captain (ret.) Mike Gagliano and Battalion Chief Casey Phillips, the other two important members of "The Seattle Guys," have been with Phil and me from the beginning of our journey in *Air Management for the Fire Service*. They have had a profound influence on this book, whether they realize it or not. Mike and Casey listened to me talk about the Core Principles of Structural Firefighting long before I even could put them together in a coherent way, and they encouraged me to continue fleshing out my ideas. Thanks, guys! Your knowledge, professionalism, and kindness have always inspired me!

I also want to thank my good friend Mike Ciampo from the FDNY. Champ's dedication to the art and craft of structural firefighting is unsurpassed. Thanks for your insights, for teaching me so much, for always making me laugh, for all the San Diego and Baltimore HOT after-parties, and for your friendship. Thanks, Paisan.

I would be remiss if I failed to mention all the past and present authors, instructors, and luminaries of the American Fire Service who have influenced me over the years. Most of what I have learned, I've learned by reading and listening and discussing and drilling with the best and brightest fire instructors in the world. I would like to thank John Norman, Vincent Dunn, John Mittendorf, Dave Dodson, Alan Brunacini (rest in peace), Mike Ciampo and Mike Dugan and the FDNY guys, Bill Gustin, John Salka, Bobby Morris, Mike and Anne Gagliano, Anthony Avillo, Francis Brannigan, Jimmy Crawford and the Pittsburgh guys, Billy Goldfeder, Ron Moore, Eddie Buchanan, Ray McCormack, Dave McGrail, Frank Montagna, Anthony Kastros, Steve Kerber, Doug Mitchell, Dan Shaw, P. J. Norwood, Forest Reeder, Danny Stratton, Cynthia Ross Tustin, Dan Sheridan, Larry McCormack, Dan Madrzykowski, Kristina Kreutzer, John Lewis, Steve Marsar, Jack Murphy, Frank Ricci, Rick Lasky, Bobby Halton, Raul Angulo, Robert Bingham, Skip Coleman, Glenn Corbett, Aaron Fields, Steve Crothers, Paul Combs, William Shouldis, Rob Schnepp, Mike Nasta, Gerry Naylis, Ed Hadfield, and a bunch of instructors I know I have not listed. These instructors had a profound effect on me and on my structural firefighting education, and I owe them a debt of gratitude I can never repay.

Tom Brennan of the FDNY deserves a special mention here. For those younger firefighters who have never heard of him, let me explain. Tom Brennan went to countless structure fires in New York City as the captain of Truck 111, one of the busiest trucks in the FDNY during the "War Years." Tom understood structural firefighting perhaps better than anyone in the last century. His monthly article on the back page of *Fire Engineering* magazine, entitled "Random Thoughts," taught countless firefighters the ins-and-outs of structural firefighting tactics. Tom holds a special place in my heart because he took the time to stop and listen—really listen—to my ideas, and he encouraged me to write them down. Thanks for all the help, Tom! Rest in peace, Brother!

Bobby Halton, the current vice president and editor of *Fire Engineering* magazine, has always been an inspiration to me. Bobby writes unrivaled, tell-it-like-it-is editorials. He is a top-shelf instructor and lecturer. He has a deep knowledge of history, which he uses to remind the fire service where we have been and where we are going. And most importantly, Bobby is dedicated to firefighters everywhere. Bobby has always supported me and my ideas. Over the years, his example and support have pushed me to be a better firefighter and instructor. Thanks, Bobby!

Diane Rothschild, the executive editor of *Fire Engineering* magazine and conference director of FDIC, deserves a big "thanks" from me as well. She published The Seattle Guys' first Air Management articles in *Fire Engineering* magazine and gave us our start at FDIC. Over the years, Diane has helped me in countless ways, from editing articles to giving advice on classroom lectures. She is a true professional. I have always valued her opinions and her friendship.

To Mary Jane Dittmar, Pete Prochilo, Ginger Mendolia, the AV crew at Indy, and everyone at *Fire Engineering* magazine and FDIC—thank you for your kindness and help over the years.

To all the crews I've worked with over the years during my career in the Seattle Fire Department—thanks for the hard work, the good times, and the memories! And to the current crews who work with me at Station 38—thanks for your professionalism, your dedication, and your commitment to making Station 38 one of the best fire stations in the city of Seattle!

INTRODUCTION

Aggressive firefighting in and around buildings of any size has always been a difficult and dangerous job. Each year, firefighters get seriously injured or even perish both inside and outside structures while they stretch hoselines, throw ground ladders, conduct searches, carry out rescue operations, and ventilate these buildings. We know how these firefighters die—usually by cardiac events, asphyxiation (running out of air), collapse injures, or thermal insult.

But how do firefighters get into trouble at structure fires in the first place? What basic knowledge or skills did they lack? What critical decisions did they make? What did they do, or fail to do, that got them into trouble? What vital information did they overlook, ignore, or not know that caused them to get injured or perish inside or outside a building on fire?

It is my belief that firefighters get hurt or die at structure fires because they lack the basic knowledge and/or comprehensive training in one or more of the core principles of structural firefighting. To safely and effectively rescue victims and extinguish fires in buildings, fire officers and firefighters must have a thorough understanding of these core principles and must know how to recognize and apply them on the fireground.

The **core principles** of structural firefighting are fire behavior, building construction, strategy, tactics, safety, and training. These six core principles of structural firefighting are best understood by looking at figure I–1.

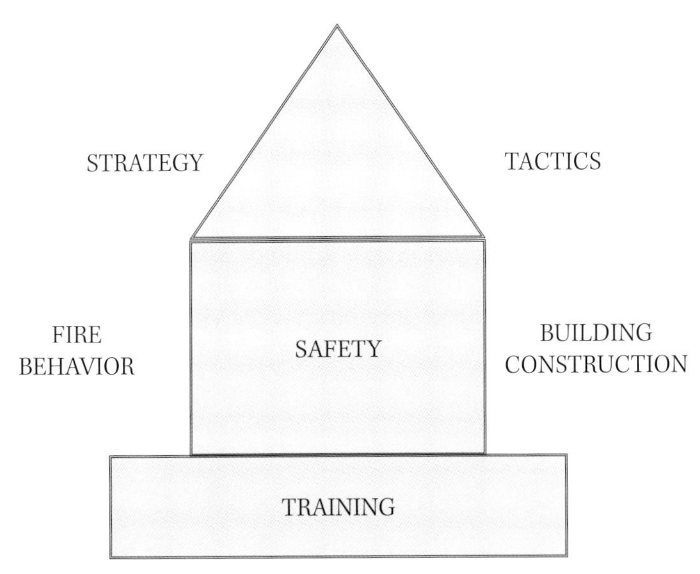

Figure I–1. Core principles of structural firefighting

Like the components of a stick-frame house, all the core principles of structural firefighting are interconnected and depend upon one another. Remove any one of the core principles, and the structure fails.

As seen in figure I–1, **training** is the foundation upon which the other five core principles stand. Meaningful and thorough training on fire behavior, building construction, strategy, tactics, and safety must start in recruit school and continue until the day a firefighter retires. In these days of tight budgets and over-scheduling, a true organizational commitment to relevant, ongoing training in these core principles is perhaps the biggest challenge that modern fire departments and fire districts face today.

In the following pages, I will address each of the core principles of structural firefighting and explain how each one relates to the others. Hopefully, by the end, you will come away with a better understanding of how these core principles must inform the critical decisions fire officers and firefighters make on the fireground. It is my belief that good critical decisions—decisions that are based upon the six core principles of firefighting—result in fireground operations that are both safe and effective. Safe and effective fireground operations mean fewer firefighter injuries and fatalities.

A Warning

I must warn the reader here. This brief study only touches upon the very basics of the art and science of fighting fires in and around structures. Under no circumstances is the reader to believe that this book is the totality of the knowledge they will need to safely and successfully fight structure fires. In fact, this book should only serve as a beginning, a starting point, to a career-long study of the core principles of firefighting. Structural firefighters and officers of every rank must never stop learning all there is to know about the core principles of this wonderful and dangerous profession. The minute any firefighter or fire officer believes that they know everything there is to know in this job is the minute that they become a liability on the fireground. When it comes to fighting structure fires, complacency kills.

Why Read About the Core Principles of Structural Firefighting?

I wrote this book to serve as a reference for firefighters of every rank who work inside or around structure fires—think of it as a primer that outlines elementary principles. This book's main purpose is to help you understand your job. This understanding, I believe, necessitates an intimate knowledge of the core principles that underlie the art and craft of fighting fires in buildings.

This brief study is organized in a simple manner. Each of the first six chapters covers one of the six core principles of structural firefighting. The remaining chapters all relate back, in one way or another, to those first six chapters. Lastly, the reader will find a bibliography and list of recommended reading. These books and articles serve as a springboard for further investigation into the world of structural firefighting.

Fire Behavior

The first core principle of structural firefighting is **fire behavior** (fig.1–1). Most structural firefighters had some type of introduction to fire behavior, sometimes called "Fire Dynamics," in probie/recruit school. To a greater or lesser extent, firefighters learned that fires in structures progress through distinct stages. Years ago, firefighters like me were taught that there were three distinct stages of a fire inside a structure: the incipient stage, the free-burning stage, and the smoldering stage. Today our understanding of the science of fire behavior has advanced, and now young firefighters are learning that there are four stages of fire development: the incipient stage, the growth stage, the fully developed stage, and the decay stage.

Figure 1–1. Fire behavior is the first core principle of structural firefighting.

Firefighters must have a clear understanding of these four stages as they relate to fires in structures. The methods of fire attack may differ depending on the particular stage the hose team encounters when they enter the structure to attack the fire. Each stage of fire development also poses its own dangers to firefighters who do not understand the risks they are up against—risks such as flashover and backdraft.

The Incipient Stage

This is the beginning of the fire event, when the fire is just starting—some call this stage the ignition stage.

Ignition happens when an ignition source (**heat**) is placed near a combustible material, like paper. This heat starts to liberate combustible gases (**fuel**) from the paper as the paper undergoes pyrolysis, a **chemical reaction** which is simply the chemical decomposition of the paper. For ignition to take place, there must be enough **oxygen** present in the surrounding atmosphere. Fire is simply a self-sustaining rapid chemical reaction that requires heat, fuel, and oxygen. In fire science classes, we learn that fire is best represented by the four-sided fire tetrahedron. The fire tetrahedron's sides are made up of heat, fuel, oxygen, and a chemical chain reaction (fig. 1–2). Take any of these sides away, and fire is not possible.

When a solid or liquid combustible material heats up, it begins to liberate combustible gases due to the chemical reaction, and it is these gases that burn. Fire is the flame and the heat that results from the burning of the combustible gases that were released during the chemical reaction. If there is enough fuel and oxygen present, the heat will cause the chemical reaction to accelerate, generating more combustible gases, which then increase the fire's growth and intensity.

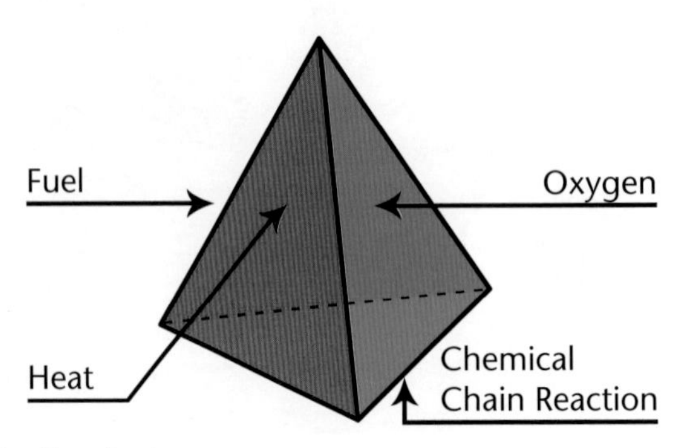

Figure 1–2. Fire Tetrahedron

If the ignitable vapor is already present, then ignition occurs almost instantaneously—think of a match being introduced to a pool of gasoline or to a room full of natural gas. If the fuel is a solid, like paper, it will take the heat from the ignition source a few seconds to produce combustible gases from the solid. Solid and liquid fuels must vaporize (release combustible gases) to burn. However, once the combustible gases begin to burn, the fire has started. Now more combustible gases are released at an increased rate due to the heat energy being produced from the fire, and the fire begins to grow.

So let me stop my brief discussion of fire chemistry here. There are many books and articles that delve much deeper into the chemistry and physics of fire, and I encourage everyone to search them out and study them.

I would like to bring our conversation back to an incipient stage fire inside a structure. At this point, there is light smoke showing and low heat, and the fire is contained to the area where it started in the room of origin. These incipient stage fires are usually easy to handle with just a pump can. However, firefighters must not be fooled. Incipient stage fires can still be dangerous, and all firefighting best-practices should be followed.

The first-arriving engine company should stretch a 1¾" handline into the fire area as a precaution in case it is needed, just as they would for a more developed fire. Even though this small fire might well be extinguished with a pump can, the hose team must be vigilant. An incipient stage fire could explosively ignite cooking oils near a stove or any aerosols that are typically found inside houses and apartments. Fire in the incipient stage can also find its way into structural void spaces through an opening or hole in the plaster or the sheetrock and move into the growth stage undetected by occupants or people passing by. Because of these and other dangers, firefighters must wear their full personal protective equipment (PPE)—bunking gear, hood, gloves, and helmet—and they should be breathing air from their SCBA. Firefighters should never breathe smoke—EVER!

I remember going to an automatic fire alarm on the eighth floor of a large apartment house in Seattle one evening. I was a lieutenant on a busy downtown ladder company at the time, Ladder 10. My partner Schon and I had taken the elevator up to floor 7, and walked up to the eighth floor to investigate an odor of food on the stove. We brought some tools with us, tools we carried to every alarm: the irons (the Halligan and flathead axe), the TIC (Thermal Imaging Camera), a pump can, and a four-foot hook. We knocked on the door of the apartment, fully expecting to find the tenant to have burnt some food on the stove, which had tripped the smoke detector and set off the building's fire alarm system. Food on the stove was our typical call to this apartment house at dinner time. But this night, no one answered the door when we knocked, so we forced the door with the irons. Immediately smoke began coming out of the upper third of the door frame, with a slight amount of pressure behind it. We hadn't put our SCBA face pieces on—a big mistake. Instead, we strode into the apartment towards the kitchen and the source of the smoke. There was a large pot on fire on the stovetop that was giving off a lot of smoke. I was holding my breath because I didn't want to breathe in the noxious smoke coming from the pot, but I couldn't hold my breath that long, and soon I had to take a breath. Immediately, I began to cough. Schon, who started coughing as well, quickly put out the fire on the stove with the pump can. Then I grabbed the smoking pot with my gloved hand, placed it in the sink, and turned on the faucet. It was then I noticed that the pot had begun to melt on the stovetop.

Schon and I, still coughing when we left the apartment, laughed and said that we were lucky that the fire wasn't more than it was—just a pot on fire on the stove. And we both said that we wished we had used our SCBAs.

I learned a valuable lesson that night. Never be complacent, and never let your guard down. I had exposed myself and another firefighter under my command to all the products of combustion for no good reason. My complacency and laziness had put us in danger at an incipient stage fire. We breathed in all the products of combustion: CO, hydrogen cyanide, benzene, ammonia, phosgene, and countless other toxins and carcinogens. What I should have done was immediately close the apartment door in the hallway once we forced it and saw smoke, and had us put on our SCBA face pieces and begin breathing air. DON'T BREATHE SMOKE. . . . It was a mistake I would never make again.

The Growth Stage

In the growth stage, the fire has enough oxygen to sustain flames. These flames generate heat, and the heat generated causes heat transfer to spread to the surrounding combustibles. These surrounding combustibles begin to off-gas sufficient flammable gases, and they soon ignite, causing the fire to grow by spreading the chemical chain reaction to nearby combustibles. Now the fire is increasing in size.

Fire growth is dependent on four variables: available oxygen, available fuel, compartment size, and the insulating value of the compartment walls and ceiling.

If the amount of available oxygen in the fire compartment is limited, the fire will have difficulty burning. In fact, if there is not enough oxygen present, the fire will not be able to sustain itself and will be extinguished. Basically, the fire is starved of oxygen and dies.

If there is available fuel, the fire will grow according to that fuel's heat release rate. The higher the heat release rate, the faster the fire will grow. The heat release rate of the fuel is the key to understanding how fast a fire will grow and how hazardous it will be for the firefighters tasked with putting it out. The faster the fuel releases its heat energy, the more severe and dangerous the fire is. A fire in a room involving three polyurethane sofas is much hotter and releases its heat many times faster than a fire in a room involving three wood benches. *Fire growth is all about the fuel's heat release rate.* If the fire does not have sufficient fuel, or if it uses all the available fuel, the fire will go out.

There are other factors related to fuels that determine the fire's ability to consume the fuel, such as the quantity of the fuel, the class of fuel (combustible solids, flammable liquids, flammable gases, and flammable metals), the quality of the fuel (wet or dry), and the density of the fuel.

If the compartment size where the fire has started is large, the heat generated by the fire will be dispersed throughout this large space and make it hard for other combustible fuels to be preheated through heat transfer. Conversely, if the compartment size is small, heat from the fire can be radiated back to the fire, allowing it to grow larger and faster.

The better the insulating value of the compartment where the fire starts, the less heat will pass through the walls and ceiling. Instead, this heat will be radiated back to the fire and any combustibles in the compartment, and the fire will grow. If the insulating value of the compartment is low, heat will pass through the walls and ceiling where the fire started, and this lost heat will not be able to help the fire grow larger.

As the fire grows, heated fire gases rise to the ceiling of the fire compartment due to convection and then begin to bank down to the floor. Some of the heat generated from the fire gases is absorbed by the ceiling and walls of the compartment, which allows some of these fire gases to cool a bit. These cooler gases are forced downward to the floor by the hotter gases that continue to rise from the fire to the ceiling level. This stratification of fire gases inside a fire compartment is called *thermal layering*, where the hottest fire gases are found at the ceiling, and the coolest fire gases are found at the floor level. Firefighters often call this phenomenon *thermal balance*.

The growing fire continues to generate more and more heated gases and radiant heat as it consumes the available combustible fuel in the room, increasing the relative pressure in the fire compartment. During this pre-flashover fire growth, more and more heat is produced, rapidly increasing the temperature in the fire compartment. Unchecked, the fire will soon flash over and become fully developed.

At this point in the fire, an attack line (either 1¾" or 2½", depending on the size of the compartment) is necessary to cool the fire compartment and get water on the fire. Firefighters must be aggressive but cautious at this stage of the fire. Heat is rapidly building up on the ceiling of the fire room, and if this heat is not cooled by a hose stream, a flashover will certainly take place sooner rather than later. If firefighters cannot find the seat of the fire in a timely manner, and hot, heavy smoke is being produced and filling up the structure, then firefighters must cool the overhead to keep this hot smoke, which is fuel, from igniting over their heads.

A "sweep up top and move" technique can be used with the hose stream, wherein the nozzle firefighter uses a straight stream and sweeps the overhead in front of him as he moves towards the seat of the fire. The water from the hoseline cools the overhead and keeps the temperature well below the flashover threshold.

The Swedish Fire Service was the first to experiment with putting quick spurts of water into the overhead during the 1980s, after a flashover in a structure killed two Swedish firefighters. Krister Giselsson and Mats Rosander were fire protection engineers at the Swedish Fire School in Stockholm, Sweden, at the time. Together, they began studying the science of fire behavior and how flashovers occurred in structure fires. They soon developed the groundbreaking technique of cooling the fire gases in structure fires by applying a mist of small droplets of water overhead with a fog nozzle. They learned that quick bursts of water applied in a tight fog pattern overhead avoided massive steam expansions, allowing firefighters a safe and comfortable environment in which to advance to the seat of the fire. This technique was later called "penciling," what we call "flow and move" today. Giselsson and Rosander understood that keeping the fire gases—the fuel found in smoke—cool stopped flashover from happening. I will talk more about Giselsson and Rosander later in Chapter 5.

Paul Grimwood of the London Fire Brigade has also written extensively on this idea of hitting the overhead of the pre-flashover fire compartment.[1] His work on three dimensional (3D) water-fog applications in structural firefighting has advanced our knowledge of flashover prevention and firefighter safety.

The modern "penciling" or "flow and move" technique can be used with a fog nozzle brought down into a straight-stream pattern, or with a smooth-bore nozzle. Some in the American Fire Service call this "sweeping the overhead," or "sweeping," or "hit and move," or "sweep up top and move." Again, applying water from an attack line overhead in the growth, or pre-flashover, phase of the fire is something every firefighter must understand and have practiced during hands-on drills. This is a basic firefighting technique that might someday save your life.

I remember a fire I had when I was a new firefighter on a busy engine company in the north end of the city, Engine 31. An arsonist was lighting fires all over our district. He had already killed several residents of a retirement home with an early morning fire he started in a common hallway.

It was 2 AM on a hot summer night, and we were dispatched to a reported house fire in our district. When we arrived we found heavy, black smoke issuing from the D-side windows and front door of a one and a half story house. My captain did a 360° walk-around and came back and told us that the fire looked to be toward the back of the house, near the kitchen. He told me to grab the 1¾" preconnect and told us to go in through the front door. All the occupants were out of the house thanks to the working smoke detectors, and the most direct route to the fire was through the front door. The Captain looked right at my partner, Tim, a twenty-year veteran, and said, "Timmy, that smoke looks hot. Make sure you cool it down so it doesn't flash."

"I'm all over it, boss," Tim said.

I pulled the 1¾" preconnect and bled the line at the front door while Tim laid down at the threshold of the front door and looked under the smoke, which was about one foot off the floor. Tim said he could see a glow toward the back of the house. We masked up, started breathing air from our SCBAs, and began to advance the line, with me on the nozzle and him backing me up. The truck company was just arriving as we started to head into the house. I remember thinking that it was hot as I crawled and advanced the hose into the living room and back toward the kitchen. I could feel the heat on the back of my neck and on my ears (this was before we had hoods). As I was making my way toward the seat of the fire, Tim stopped me and said that it was getting too hot. He told me to hit the overhead with the stream, something he had coached me on in drills back at the station. I opened the nozzle in a straight stream pattern at the ceiling and swept it for about five seconds or so. No water came back down ahead of us, so Tim told me to hit the overhead a few more times, which I did. On the third or fourth burst, I could hear drops of water coming down in front of me.

"That's it," Tim said in my ear. "It's cooled down, let's go." We pushed on into the kitchen, which was completely on fire, from the cabinets to the kitchen table. I knocked it down quickly from the living room entrance to the kitchen, just as the truck company was starting up the PPV fan behind us. Soon the smoke had lifted. It was a lot cooler in the living room now, and I could see everything clearly. We advanced into the kitchen and soon had the fire under control.

It turns out that the fire had been set by the arsonist, who had thrown a Molotov cocktail into the kitchen through an open window while the occupants of the house were asleep. Luckily, no one had been killed or injured, including us. But I will never forget what Tim taught me that night. Under Tim's instruction, I had cooled the overhead in front of us and kept the fire from rolling over and flashing over on us as we advanced to the seat of the fire. It was a technique that I would use for the rest of my career.

Flashover

Flashover is an event that occurs at the end of the growth stage of a fire. During the growth stage of a fire, heat is produced inside the fire compartment. Some of this heat is present in the heated smoke and fire gases that the fire is producing, which rise to ceiling level and then bank down to floor level. Some of this heat is radiated back from the walls of the compartment and transferred to the contents of the room. The temperature of the contents rises, which causes these contents to begin off-gassing ignitable vapors, which are nothing more than unburned fuels. The ability of the contents to absorb the transferred heat of the fire is determined by many factors, such as their physical states and their ignition temperatures. A fire in a compartment can produce very high heat levels (upwards of 2000°F). The quickly rising heat in the compartment causes all the contents to give off ignitable gases, which firefighters recognize as smoke. This smoke is typically hot, quickly boiling out of an opening (like a door or a window), thick or dense in appearance, and black.

Flashover occurs when all the fuel present in the compartment—all the accumulated ignitable smoke, gases, and the gases being released from the exposed combustible surfaces of all the contents—ignites suddenly. Put another way, flashover occurs when all the surfaces exposed to thermal radiation reach ignition temperature simultaneously and fire spreads rapidly throughout the space, resulting in full room involvement or total involvement of the compartment or enclosed area.

Fire needs oxygen for the gases to ignite. Typically, the fire gases being produced are too rich to burn inside a closed compartment like a room. But when a firefighter opens the front door of a house or ventilates several windows of the fire room, oxygen is introduced into the space, and flashover can occur rapidly.

A flashover is sudden, momentary event that can take place in seconds, or even in a fraction of a second. The temperature of the fire compartment increases instantly to well over 1800°F. These conditions are immediately untenable for firefighters, and will result in serious injury or death to any firefighter in the fire compartment within seconds.

The warning signs of a potential flashover are the following:

- The buildup or accumulation of heat and smoke within a structure or room
- Smoke that is increasing in temperature (heat), volume, velocity, density, and color (we have Dave Dodson and his *Art of Reading Smoke* program to thank for these terms, which have become part of the common language firefighters use when discussing smoke)
- Firefighters being forced to stay low in the fire compartment due to increased heat
- Flames visible in the overhead smoke, or *rollover*

If you use the Thermal Imaging Camera (TIC) to look up at the ceiling into the black smoke, oftentimes you will see flames rolling along the ceiling, flames you cannot see with the naked eye. Rollover occurs when heated fire gases rise up to the ceiling of the fire compartment during fire growth, leading up to the flashover stage of the fire. These fire gases and smoke are fuel, and given the right conditions, will ignite. Flames appear at the ceiling level and will travel with the smoke to areas of less pressure. This rollover phenomenon is sometimes called *flameover.* The danger with rollover is that fire can travel over firefighters' heads and get behind them, blocking their exit pathway. Rollover also allows the fire to spread to areas in the structure with less pressure, making it appear that there is more than one point of fire origin. Think of a fire that started in the kitchen of a ranch-style house, but due to rollover, flames are exiting the house from an open back bedroom window.

Every firefighter must know these warning signs of flashover, the deadliest of fire events.

The Fully Developed Stage

The fire has reached the fully developed stage when the fire has flashed over and now involves the entire compartment or space of a structure, and all the contents are burning. The rate at which the fully developed compartment fire is burning is limited by the amount of ventilation, or oxygen, the compartment is receiving. Picture a fire in a wood burning stove. The wood in the stove burns faster with the damper wide open than with the damper closed. The same concept is at work in a fully developed compartment fire. The more oxygen the fully developed fire receives, the faster the rate of burning in the compartment. This type of fully developed fire is called a *ventilation-limited* fire.

Because of the excellent fire dynamics work of Stephen Kerber and the team of dedicated scientists at the UL Firefighter Safety Research Institute, we now know that all fires in structures that have reached the fully developed stage are ventilation-limited fires. Even if you see flames issuing from an open window, the fully developed fire inside the room is ventilation-limited. This fire is starved for oxygen, and when it finds the right mixture of oxygen, it will ignite (fig. 1–3).

These fires usually occur in insulated buildings, such as houses, apartment houses, and commercial structures. Fully developed, ventilation-limited fires typically have large amounts of smoke moving out of the structure rapidly under pressure. There is not enough oxygen to sustain full ignition of all or most of the fire gases being produced.

During a ventilation-limited fire, ceiling temperatures in the fire room are high due to the rapid accumulation of smoke and hot fire gases in the compartment. These temperatures are typically 1100°F–1500°F, and may be even hotter depending on what materials are burning in the fire room. Since large amounts of smoke are

Figure 1–3. Multiple engine companies lay 2½" attack lines at a fire in a commercial occupancy.

common in ventilation-limited fires, we must always remember what Dave Dodson has told us for years, that **SMOKE IS FUEL**. The smoke in the fully developed stage of a fire is a mixture of particulates and unburned fire gases, and these fuels all have different ignition temperatures and flammability ranges. Dodson calls them "ladder fuels," and structural firefighters need to know that whenever they are crawling through smoke, they are surrounded by, and moving through, unburned fuels. Given the right heat/air/fuel mixture, these fuels can ignite catastrophically.

Sometimes a fully developed fire has plenty of ventilation (oxygen) and is only limited by the amount of fuel available to it—think of burning pallets out on a driveway, or a fire that has burned through the attic space and roof of a single-family dwelling and is now free burning. This type of fully developed fire is called a *fuel-limited* fire. This type of fire receives plenty of oxygen. This fully developed fire is only limited by the amount of fuel available to it. Once all the fuel has been burned by the fire, the fire will die out. We see fuel-controlled fires only in structures that are well ventilated—which usually means that they are fully involved. In other words, the structure itself is on fire, not just a room or two, or three.

Every fully developed stage fire that occurs inside a structure is likely ventilation-limited, unless several exterior walls of the structure have collapsed or have been consumed by fire.

Backdraft

A fire in a structure needs oxygen to burn. When a fire burns in a closed or confined area, such as a room in a structure, the fire will consume the available oxygen and produce large amounts of fire gases and heat. As the smoke and fire gases are heated, they expand, pressurizing the fire compartment, and force smoke out from all available openings. As the oxygen level drops, visible flames begin to decrease—the fire is being starved of oxygen, which is necessary for continued combustion. The fire compartment is now filled with high heat (typically over 1300°F), along with smoke and super-heated fire gases (fuel). One of these fire gases, carbon monoxide, has a particularly large flammable range—from 12.5 percent to 74 percent. The fire now has heat and fuel, and only needs oxygen to ignite explosively. When oxygen is reintroduced to this environment, the smoke and fire gases ignite in a violently explosive manner. This backdraft has the power to blow out windows, doors, and bearing walls, and to seriously injure or kill unsuspecting firefighters.

The warning signs of a potential backdraft are the following:

- Any fire in an unventilated or under-ventilated compartment or space
- Large amounts of dark grey or yellow-grey smoke issuing from the structure under pressure
- Doors and windows that are hot to the touch
- The smoke coming from the structure may appear turbulent or boiling
- The smoke coming from the structure may be observed, at times, to be drawn back into the building, as if the building is breathing

The keys to preventing a backdraft from occurring on the fireground are recognizing the warning signs and either performing vertical ventilation directly over the involved area, or introducing enough water into the space to cool the superheated gases.

For years, Chief (ret.) John Mittendorf has explained that vertical ventilation directly over a potential backdraft will reduce both the internal temperature and the pressure inside the fire compartment by exhausting superheated smoke and fire gases to the exterior of the structure.

If vertical ventilation is not feasible or even possible, the only other choice is to overwhelm the fire compartment with water from a safe distance, using 2½" hoselines, monitors, and deck guns. This is a defensive strategy that may cause the backdraft to occur when water is forced through an intact window or doorway. Since everyone on the fireground will know, the impending backdraft will not surprise any firefighter.

Firefighters See More Flashovers at Today's Structure Fires

Today's firefighters encounter flashovers much more frequently than they do backdrafts. Why is this?

In these modern times, with automated fire detection systems, instantaneous communications, and computer aided dispatching, fire companies arrive at structure fires quickly, when the fire is still contained to the room or rooms of origin. Advanced fire detection systems and shorter response times allow firefighters to get into these structures fast, before these fires have reached flashover and are fully developed. Today's fires are also burning hotter. The plastics and synthetics used in home and office furnishings give off more BTUs (or kilowatts of energy) and release this heat faster due to the very fast *heat release rates* of these materials. Modern structures are well insulated and tight (think double and triple paned windows and R36 insulation in the walls), so flashover becomes a real and common danger for today's structural firefighters.

The reality for firefighters today is that there is a new time/temperature curve for modern, synthetic contents burning in a compartment or structure (fig. 1–4). The high temperatures and the fast heat release rates of burning synthetic materials allow fire in a structure to quickly fill up the open space where the fire has originated with hot smoke (or fuel) that is too rich to completely ignite into flames. The temperatures of this superheated smoke can range from 400°F to well over 1200°F. When firefighters arrive at a room fire that is consuming synthetic materials and then either break a window to ventilate the fire, or open the door to the space to attack the fire, they allow fresh air to flow into the fire room, which supplies oxygen to the fuel-rich smoke. These firefighters have introduced a new flow path of air to the fire. **A flow path is the volume, composed of area(s) within a structure, between an inlet and an outlet that allows the movement of heat and smoke from the higher pressure fire area towards the lower pressure areas by way of doorways, stairways, or windows.**

The air flow provided by the intentional or unintentional ventilation by the firefighers quickly provides oxygen to the starved fire and a flashover occurs.

A fully developed fire burning inside a structure is the type of fire that firefighters dream about fighting. The fire location is obvious, due to the visible open flaming of the burning material. The engine company that arrives first at a fire

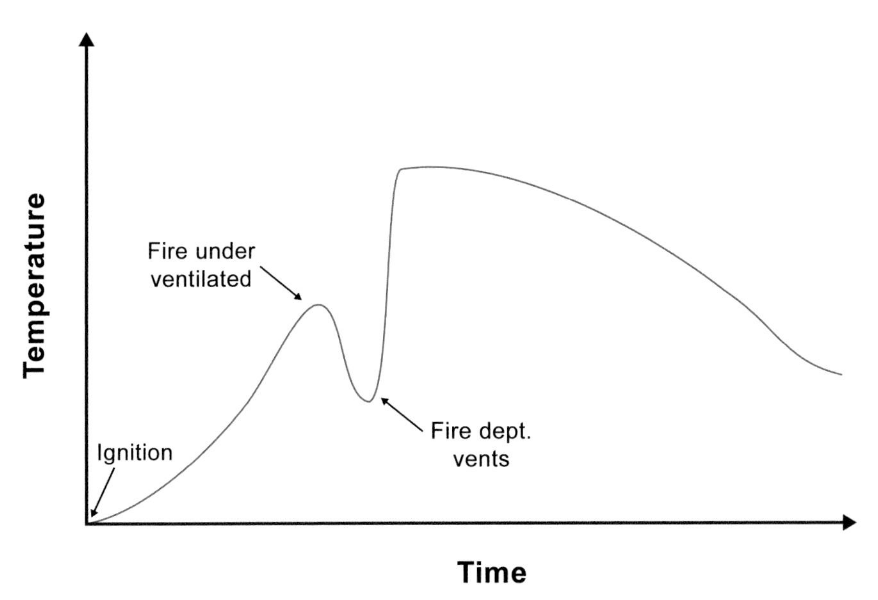

Figure 1–4. New Time/Temperature Curve (Courtesy of UL and Steve Kerber)

in the fully developed stage understands that they have a working fire, a fire that they will attack either offensively or defensively, depending on the size of the fire, the size of the structure, and the risk/benefit analysis.

Observing and Making the Right Decisions

I recall arriving first at a fully developed room and contents fire at a house one night. I was working overtime on Engine 28, a busy, top-notch engine company at the south end of the city. We were returning to the station after having just put out a car fire when we were dispatched to a report of a house fire not far away. We already had our bunking gear on, so we scanned the night sky in the direction of the cross-streets the dispatcher gave us.

"I see smoke," said Murph, my partner for the shift. At that time, Murph had been in the fire department for over twenty-five years and had served in Vietnam. "We've got a worker," he smiled.

As we pulled into the block, I could smell that tell-tale house fire smell. The driver pulled past the house on fire so that the lieutenant could see three sides while leaving the street in front of the house open for Ladder 12, which was responding from Station 28 and was a few minutes behind us. As we pulled past, I saw heavy flames coming out of three windows on floor two of the house and thick smoke coming from the eves of the roof.

When the rig stopped, the lieutenant did a quick walk-around of the house. "Pull the 1¾" preconnect up to floor 2," he told us. "I don't see anything in the back. No basement. No cars in the driveway, no visible victims. Take a pike pole to pull the ceilings once you get the room fires out on floor 2, because it's gotten in the attic."

Murph grabbed a pike pole off the rig and went ahead of me to do a quick recon while I grabbed the hose bundle off the rig. I began flaking the hose behind me and headed up the walkway in the front yard.

"Front door," Murph said, returning from his recon. "The stairs are right inside to the left. No smoke on floor 1. The fire on floor 2 has already vented." He started helping me lay the hose out in the front yard so that we would have an easy pull in the front door and up the stairs to the fire. By taking the time to do a good recon and to lay out the hose correctly in the front yard, we were setting ourselves up for success in getting to the fire.

Meanwhile, the driver had already gotten a hydrant supply and was ready to charge the preconnect for us. I signaled for him to charge the line, and Murph and I masked up and began breathing air. I put on my gloves, bled the line, turned on my flashlight, and then moved quickly through the front door and up the stairs. The first floor was clear—no smoke. When I hit the top of the stairs, I saw smoke halfway down the walls of the hallway and fire traveling out of two rooms and heading toward us on the ceiling. One of the doors to the rooms was open, and the other appeared to be burned through at the top.

I opened up the nozzle from the top of the stairway and immediately hit the hallway ceiling where fire was rolling over. This cooled the overhead and pushed the flames back into the two rooms—I was cooling the overhead from a relatively safe location. I closed the nozzle and began moving down the hall towards the fire. Murph was at the top of the stairs, feeding me hose, as I crawled down the hall to the first room. I positioned myself outside of the first door and saw that the entire room was involved in flames. I opened the nozzle, rotated it several times in circular motion to cover the room, and knocked the fire down quickly.

Murph had made his way up to my position, and I heard him say in my ear, "Nice work, we'll finish it up after we get the other room."

We quickly made our way to the other room, Murph making sure I had enough hose to do the job. Again, I positioned myself outside the fire room door and opened the nozzle, hitting the entire room, ceiling to floor in several circular motions. The main body of the fire in the second room was extinguished quickly.

"Command from Engine 28 Team B, we have water on the fire on floor 2," I heard Murph say over the radio.

As we were making our way into the second room to hit the small spot fires, I could hear Ladder 12's chainsaws on the roof.

"Command from Ladder 12, we have a good hole open in the roof, with heavy flames visible," I head the Ladder 12 officer report over the radio. "We need a protective hoseline up here."

Murph immediately began to pull ceilings in the second fire room. He had the entire ceiling down in a very short time. The attic was fully involved in fire. I opened the nozzle and hit the attic space, moving the stream over the underside to the sheeting until all I could see was steam. The fire was out quickly. Soon I could see the guys from Ladder 12 working above me, making their vent hole bigger. The second truck company, Ladder 3, was finishing up their primary search on floor 2.

"Command from Engine 28 Team B, we have water on the fire in the attic," Murph reported. "The fire is under control."

I looked back down the hallway and saw that the crew from Engine 33 was in the first fire room with a hoseline and was pulling ceilings. Their officer came over the radio and said that the fire in their location was under control and they were checking for extension.

I patted Murph on the back, and we checked our air. We each had a little less than half a bottle left, so we began pulling some sheetrock in the room to make sure that there was no hidden fire. Soon another engine company came up to relieve us, and we made our way out of the house.

As we made our way back to our rig, I couldn't help but think of how everyone on our engine company had made good observations and decisions. Our officer did a 360° walk-around which allowed him to make an accurate initial size-up of the fire. Murph and I took the time to flake out the hoseline so that we would have an easy pull into the house. Our driver found a hydrant and had a supply. Murph did a quick recon of the house so that we knew exactly what we were up against. We recognized that the room fires on floor 2 had already vented out the windows and that the fire was in the fully developed stage. We selected the right tool to take with us to pull ceilings so we could attack the fire in the attic. Murph and I worked together as a team to get the hoseline up to floor 2. I hit the fire in both the rooms from the relative safe location of the hallway. And Engine 33 was right behind us with the backup line in case anything went wrong.

It was one of those fires where everything went right. Everyone knew their job, everyone did what they were supposed to do, and there were no surprises. Unfortunately, many fire operations in structures don't always go this well. Back at the rig, we all smiled at one another, joking around, giving each other a hard time—the way all firefighters do when the fire is out—and wishing like hell that we could do it all over again.

The Decay Stage

When the fuel in the compartment or space has been mostly consumed by the fire, the fire will decrease in size and intensity, and the heat released by the fire will decrease significantly. At this point, the fire is in the decay stage; the fire is not burning freely anymore. Glowing embers are a sign that the fire has expended almost all the contents of the room or compartment—there is simply no more fuel to burn. Given enough time, the fire will self-extinguish due to lack of fuel. However, the decay stage can also occur when a fire has run low on oxygen. This can happen even if there is plenty of fuel still in the fire compartment. Remember the four-sided fire tetrahedron! Take away the oxygen **or** the fuel, and the decay stage begins and the fire will eventually go out.

In the decay stage of the fire, there may still be small spot fires, or smoldering fires, burning in the consumed contents of the room, and these must be extinguished with a handline. It is important that fire crews "open up" the structure by pulling walls and ceilings to make sure that the fire has not made its way into the void spaces of the building. Just because a fire is in the decay stage in one portion of the structure does not mean that fire is not burning freely behind a wall or in the overhead space. Good overhaul techniques must be followed. Firefighters must make sure they extinguish the fire completely and do not leave embers burning anywhere inside the structure. All furniture in the fire room should be taken outside. All cabinets, bookcases, and moldings must be removed if they were exposed to fire or high heat. All combustibles that might still be smoldering have to be taken outside the structure and put in a debris pile well away from any buildings. This pile should be wet down before fire crews leave the scene.

Firefighter safety should not be disregarded during the overhaul of the fire. I cannot tell you how many times I have seen firefighters assigned to overhaul go into the fire building without their SCBAs on, or worse, wearing their SCBAs on their backs but not wearing their face pieces and breathing air.

The post-fire environment inside a structure is just as dangerous, and perhaps more dangerous, to firefighters as the fire environment. In fact, several important studies have shown that firefighters who do not use their SCBAs during overhaul are exposing themselves to deadly toxins, such as hydrogen cyanide (HCN), and carcinogens, such as benzene, acrolein, and formaldehyde. There is also evidence from a study out of the University of Cincinnati that firefighters who do not wear their SCBAs during overhaul are breathing in dangerous levels of ultrafine particulates, which are invisible to the naked eye and are released at every fire.[2] These ultrafine particulates are being inhaled into the deepest compartments of the lungs by unprotected firefighters during overhaul, increasing these firefighters' risk of heart disease, sudden cardiac arrest, and cancers.

Officers must be vigilant in making sure that their firefighters wear and use all their protective gear during overhaul—SCBAs, hoods, bunking gear, *and gloves.* We are just now learning that many of the carcinogens and toxins found in the post-fire environment are absorbable through the exposed skin of the hands, neck, and face. So wear your PPE during overhaul, and wash your gear and take a shower when you get back to the station. Today, many fire departments are providing their firefighters with two sets of bunking gear so that after a structure fire, firefighters can change out their dirty, contaminated bunkers for a clean set back at the station.

Cancer is the silent killer of firefighters, and our members—both active-duty and retired—are dying every week in staggering numbers around the world.

To help decrease firefighter cancer deaths, firefighters in Boston are using baby wipes at the fire scene to clean their faces, hands, and necks of the smoke particulates that they may have been exposed to during the fire or during overhaul activities. This seemingly small act, along with our increased awareness, will hopefully pay huge dividends in our fight against the plague of cancers that are decimating our firefighters.

Endnotes

1. Grimwood cites a real life reverse flashover scenario here: https://modernfirebehavior.com/flow-path-reversal-by-paul-grimwood/
2. https://www.ncbi.nlm.nih.gov/pubmed/24512044

2

Building Construction

The second core of structural firefighting is **building construction.** Structural firefighters fight fires in and around buildings. Therefore, firefighters must understand how buildings are put together, what the common structural components are that "hold up" a building, what materials are used in building construction, and what the dangers are when fires occur in buildings.

Since the methods and materials used in building construction are constantly changing, firefighters must continually study building construction throughout their careers if they want to operate safely and effectively at structure fires. As Francis L. Brannigan, the father of building construction for the fire service, preached throughout his career, "The building is your enemy. Know your enemy."

This chapter is not intended to cover every aspect of building construction—books have been written on this subject. Instead, I want to give the reader a brief overview of basic building construction styles, structural components, and connections, and to discuss the dangers associated with them when they are exposed to fire and firefighting operations.

When building architects and engineers design structures and the materials used to build them, they must consider the force of gravity, which is constantly working against the structure. Brannigan reminds us that a building is nothing more than a "gravity resistant system" that is built with the materials available at the time of construction. These buildings, or gravity resistance systems, are the sum total of all the structural elements and the connections that support and transfer the loads of the structure. Once fire begins to destroy the structural elements or the connections of the building, or both, a collapse can occur. Building collapse is one of the greatest dangers structural firefighters face while working at a fire in a building (fig. 2–1).

Figure 2–1. Building collapse is one of the greatest dangers structural firefighters face while working at a fire inside a building.

CONSTRUCTION STYLES

The vast majority of the structures that exist today in the United States are built in either one of two styles: conventional construction (sometimes referred to as "legacy construction") or lightweight construction (sometimes referred to as "modern construction").

Conventional construction utilizes structural components that rely on size for strength. The greater the distance that the structural component must span, the larger it must be. The greater the load, or weight, that the structural component must carry, the larger it must be. We see this in our older buildings. Take, for example, pre–World War II houses. Most were built using large and true dimensional lumber, such as 2×6s, 2×8s, 2×12s, 4×8s, or larger, as basic building components. These larger-sized structural components can burn for a significant amount of time before they fail due to their true dimensional size. Their strength under heat and fire conditions comes from actual size (fig. 2–2).

Today's lightweight construction does not utilize structural components that rely on their size for strength. Instead, lightweight construction's strength is derived from smaller, lighter components that are arranged geometrically for tension and compression stability—usually in a triangular configuration, such as the truss. What is a truss? A truss is a structural component using geometrically arranged triangular units to attain stability. A truss is composed of a top and a bottom chord with internal web members arranged in triangular units connecting the top and bottom chords. Typically, the top chord of a truss is in compression, while the bottom chord is in tension. However, this can be reversed in certain truss applications, such as we see in the beams of an aerial ladder.

Lightweight construction's strength depends on the geometry of its members, not size. Truss systems that are used to build a common house today are mostly made of 2×3s—found in wooden I-beams used as floor joists or flat roof joists—and 2×4s—found in peaked roof trusses and in parallel chord trusses used as flat roof joists and floor joists (fig. 2–3).

John Mittendorf has pointed out for decades that a truss is only as strong as the connections that hold it together. Oftentimes, these connectors are metal

Figure 2–2. The larger the structural components, the more time it may take for the structure to fail. (Courtesy of Mike Dugan)

Figure 2–3. Trusses are components arranged in triangular units to sustain structures.

gusset plates that have been shown to fail in a matter of minutes when exposed to the heat and flames of a well-involved fire. Interestingly, some truss suppliers are now using plastic gusset plates as connectors, and these fail even faster than the metal ones.

The smaller size of the lightweight truss components, together with the inherent weakness of the connections that hold these small components together, requires less exposure to heat and fire for structural collapse to occur than do conventional construction components.

I remember inspecting an old manufacturing building in the industrial area of south Seattle a few years back. The building was occupied by a trailer manufacturing company that made all sorts of trailers—utility trailers, boat trailers, horse trailers. The building was classified as a "heavy timber"—or class IV construction. It had exterior brick walls with huge interior wood columns and beams—12×12s and larger. As I was walking through the building with my crew and the business owner, a guy named Art, we came across a section of the building that had been involved in a fire several years earlier. The fire had been burning for quite a while before it was discovered, but fire crews from Engine 6 and Engine 30 had initiated an aggressive interior attack with 2½" lines and had put it out relatively quickly.

The 12×12 beams that had been exposed to heat and fire had heavy black char on them. In fact, some of the beams closest to the seat of the fire were "alligatored"—the charring was advanced enough to create the classic alligator skin appearance that tells firefighters it was exposed to high heat and was actively burning.

"The fire started from an electric heater that one of our guys forgot to turn off before he left for the night. Luckily someone saw the smoke and you guys put it out before the whole place burned down," Art said, pointing to the charring on one of the beams.

One of the guys on my crew that day, Bobby, said he was amazed the 12×12 beams that had the heavy charring did not have to be replaced.

"No, the building engineers who looked at them told us that they were OK, so we left them as they were, char and all."

After we left the building, I reminded Bobby and the crew that structures built using conventional construction derived their strength from the size of their building components, and that these large components could withstand quite a bit of heat and fire.

"The amount of fire that the 12×12s were exposed to would have probably caused a collapse if lightweight trusses were used as the beams instead," I told them.

They all agreed and said that they wished all buildings were constructed with conventional building components.

"Those lightweight trusses are firefighter killers," Bobby remarked. We all nodded in agreement. As we walked back to the rig, I couldn't help but notice how a simple building inspection had transformed into a fantastic drill on construction styles and firefighter safety. I promised myself that I would try to use every building inspection as another opportunity to train my crew on building construction.

FUNDAMENTAL STRUCTURAL COMPONENTS

Buildings are constructed using structural components, which are sometimes called *structural elements* or *structural members*. Firefighters must have a basic knowledge of how these individual structural components function to carry a load and transfer it to the next component in the structural framework of a building. By understanding the basics of structural components, firefighters can then recognize the associated dangers when these structural components are weakened by heat and fire (fig. 2–4).

BEAMS

A beam is a supporting component that transfers weight, usually horizontally, from one direction to its points of support. A beam is usually a long, sturdy piece of timber or metal spanning an opening or part of a building, and it is usually supported by columns, which I will discuss in a moment.

Picture a heavy weight, like a bed, placed on the floor of a second story room. The floor joists act as small beams, which receive the weight of the bed, transfer it horizontally in both directions, and deliver the weight to the opposite walls of the room where the floor joist is supported. Beams, along with columns, are probably the oldest structural components known to humankind.

Figure 2–4. The more that firefighters understand structural components, the faster they can recognize dangers associated with them on the fireground.

To understand the forces acting on a loaded beam, picture a horizontal piece of wood supported on each end by a 2'×2' concrete block. When a load, or weight, is placed upon the beam, the beam bends downward. The top of the beam shortens and is in compression, while the bottom of the beam elongates and is in tension.

When fire attacks a wood beam that is carrying a load, the fire destroys the beam's mass. As the wood of the beam burns away, the beam weakens. If the fire destroys enough of the beam, then the beam can no longer support the weight of the load, and a collapse will occur.

A *cantilever beam* is supported only at one end. Picture a balcony or an exterior walkway that projects out from a building with no vertical supports, but which is supported by its connections to the structure at one end. A wood cantilever beam exposed to heat and flames can fail in one of two ways. First, it can fail if the fire destroys enough of its mass. Second, it can fail if the fire causes its connections to the building to weaken and pull away from the building. The collapse of balconies that are supported by cantilever beams have killed or injured many firefighters. Firefighters working on or under cantilever beams at a structure fire must understand that if the beam or its connections to the structure are involved in fire, then they are in danger of being caught in a collapse.

An *I-beam* is simply a lightweight beam whose cross section is in the shape of an "I". The I-beam is composed of a top and bottom flange separated by a vertical web. I-beams are typically made of steel and used in steel frame buildings. However, lightweight wooden I-beams—also called *TGIs* (tongue-and-groove I-beams), *engineered wood joists*, and *PRIs* (performance rated I-joists)—are now commonly used as joists in both floors and ceilings of many newer wood frame structures.

Columns

A column is a structural component that transmits a compressive load along a straight path. Columns are typically vertical, but can be oriented in a more horizontal orientation—such as a raker or strut that is helping to hold up a wall or a vehicle on its side.

Vertical columns usually hold up horizontal beams that have a load, or weight, imposed on them. Picture the massive posts, or columns, holding up the large beams in an older manufacturing building of heavy timber construction.

Columns fail by buckling or by being crushed, depending on their size. Tall and thin columns tend to buckle when they fail. Short and thick columns tend to be crushed along their axis when they fail.

Arches

Arches combine the function of both a beam and a column. Arches span a space while supporting weight. All the forces on the arch are compressive—that is, the arch carries its load in compression. The ancient Romans were the first to develop the use of arches in a wide range of structures.

There are basically two types of arches: curved and flat arches. Arches fail when any part of the arch is removed or damaged, or when the adjoining wall is damaged or fails.

Walls

Walls are building components that transmit the compressive forces applied along the top, or received at any point, to the ground. A wall functions like a tall, slender column. Picture a row of beams being held up by two parallel walls. The beams transmit their load horizontally to the walls, which transmit this load to the ground along their length.

There are basically two categories of walls: load-bearing and non-load-bearing. Load-bearing walls can be either interior or exterior walls. Load-bearing walls carry a load of some part of the structure as well as the weight of the wall itself. Non-load-bearing walls support only their own weight.

Many different types of walls exist: cross walls, veneer walls, composite walls, shear walls, panel walls, curtain walls, party walls, fire walls, partition walls, demising walls, cantilever walls, parapet walls, etc. Though it is beyond the scope of this discussion, firefighters need to become familiar with all these types of walls and their individual characteristics.

Walls can fail in many ways. They can be pushed out of plumb, lose their horizontal support or bracing, buckle under load, lose their connection to the ground or other structural members, or become weakened by fire burning or heating the material of the wall itself.

Roofs

A roof's main purpose is to keep the weather—sun, snow, and rain—out of the building. However, in some cases, the roof is also necessary to the structural stability of the building—such as a tilt slab concrete building. Fire in this type of structure can damage the metal open-web bar joist trusses, which make up the roof that holds the walls in place, and can cause a collapse.

There are many different roof types: gable, flat, mansard, shed, hip, gambrel, arched, lantern, saw tooth, butterfly, etc. (fig. 2–5). All firefighters should know these roofs by their common names and understand that each roof type has its own, unique hazards. Firefighters also need to understand that all roofs in buildings built in the last thirty years or so are probably constructed using lightweight components, or trusses.

Firefighters must also be aware that sometimes buildings have roofs under roofs. That is, instead of re-roofing the existing roof of a structure, the building owner simply built a new roof over the old one. These are typically found in older

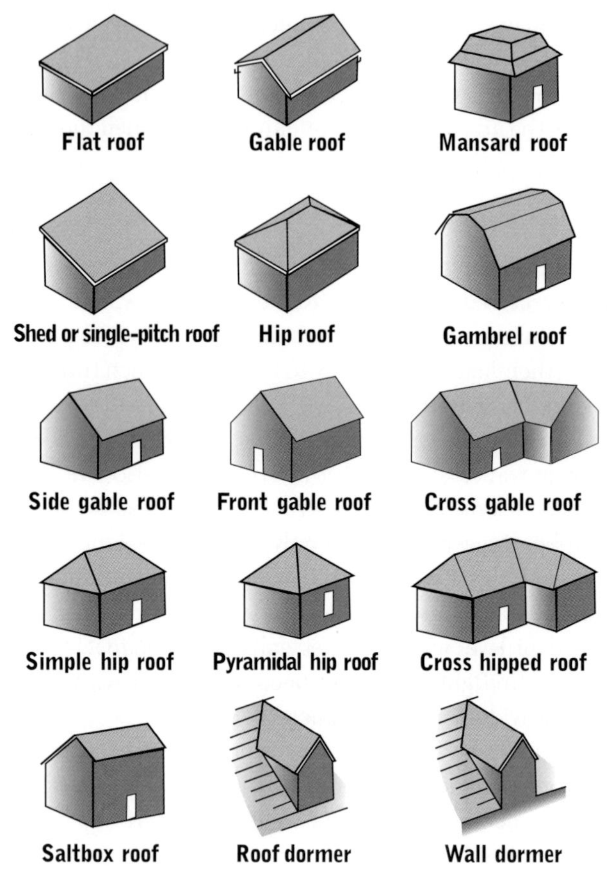

Figure 2–5. Roof Types

buildings that have had some sort of renovation done to them. Fire under the lower roof can be hidden, and it is very difficult to ventilate.

Fires in the space under roofs and above the ceiling—typically called *attic fires* in buildings with pitched roofs, or *cockloft fires* in buildings with flat roofs—are hard to ventilate and extinguish, and they can cause a collapse of the roof onto unsuspecting firefighters. Engine and ladder companies need to work in a coordinated fashion to attack and extinguish an attic or cockloft fire. If a fire department or fire district does not have a dedicated ladder company, one of the engine companies at the fire must act as a ladder company and ventilate the fire. Just because your department does not have a ladder truck does not mean that truck functions can be ignored. To do so is to invite disaster to the crews operating on the interior of the structure.

I remember when I was first transferred to downtown Seattle, to Station 25. I was the new lieutenant on Ladder 10. The crew was made up of mostly older firefighters. My driver, Ted, had twenty-eight years in, and the tillerman, Gary, had thirty. Both had been working on Ladder 10 for over twenty-five years. Compared to them, I was the new guy, the "boot," with twelve years on the job.

I knew that the crew was waiting to see how I acted at a working structure fire, to see if I knew anything, to see what type of fire lieutenant I was. They didn't have long to wait. Two weeks into my new assignment, we caught a working fire around two in the morning at a four-story apartment building in the University District, near the University of Washington. The building was brick—typical ordinary construction—about 60'×300' in size with a flat roof. Heavy flames were venting out three windows on the fourth floor, and smoke was pushing with some pressure from the cockloft. The building was occupied by mostly college students.

We were the second due ladder truck. While we were responding, we could hear the officer of Ladder 9, the first truck, talking over the radio. Ladder 9 had put their aerial to the strong side of the roof. Half the crew was inside the building, evacuating tenants and searching floor 4, while the other half was on the roof, cutting an offensive heat hole over the fire. The Ladder 9 officer on floor 4 was reporting heavy smoke down to the floor, flames visible in the hallway, and a high heat condition.

As we turned the corner into the block, I could see that there was no way we could get our rig close enough to the fire building to use our aerial. The street was narrow, there were several Engines in our way, and 4" supply hose was laid out all over the street.

"OK," I told my crew, "let's grab our SCBAs, saws, and roof hooks, and use Ladder 9's aerial to get up on the roof."

As the crew gathered up our chainsaws and hooks, I grabbed the thermal imaging camera (TIC) and confirmed with the incident commander (IC) that Ladder 10 was going to the roof to ventilate. I also requested that the next ladder company place a 45' ground ladder as a second means of egress for us.

"Follow me," I said, as I took a roof hook and started climbing up Ladder 9's aerial. About halfway up, I stopped, put on my face piece, turned on my SCBA, and began breathing air from my bottle. When I reached the roof, I saw that there were three fire walls extending through the roof by several feet, each wall about seventy feet from the next. I also saw heavy smoke at the far end of the building.

I used my roof hook to sound the roof area around the aerial ladder that I would be stepping off onto. It was solid, so I climbed off the aerial and sounded a bit more. My crew was right behind me, breathing air from their SCBAs and ready to go. I told Ted to make an inspection cut to see what type of construction we were up against, what the roof decking was made of, and what smoke conditions were like under our part of the roof.

The inspection cut told us that we were on a conventionally constructed roof, with full-sized 2×10" joists, covered with 1×4" shiplap. The shiplap was covered with tar paper and a hot-mopped tar barrier on top. We didn't see any smoke from the hole.

"OK, let's go," I said. I began walking toward the fire, sounding ahead of me with the roof hook, and my guys followed me in single file, just as they had been trained to do—Ted with a chainsaw behind me, Gary with a hook behind him, and Jrob following with the second chainsaw. We made our way over the first fire wall and cut a small smoke-indicator hole just on the other side. No smoke came out, so we continued on in single-file over the next fire wall and cut another smoke-indicator hole. No smoke. So we made our way to the last fire wall. I stopped on the strong side of the last fire wall. On the other side, about halfway down the last section of roof, I saw flames lapping up from the windows of the apartments below and from the offensive heat holes that Ladder 9 had cut. There was also heavy smoke lying over the last section of roof, and two members from Ladder 9 were making their way back to the aerial ladder, their low air warning bells ringing on the SCBAs.

Duncan from Ladder 9 grabbed my arm. "The fire is running the cockloft," he said through his mask. "We're low on air and need more saws up here." I nodded and told him that we were going to cut a defensive trench cut on this side of the fire wall to try and cut it off.

I motioned for my crew to gather around and huddle up so they could hear me. "We are going to stay on this side of the fire wall, the strong side, and cut a trench to try and cut the fire from taking the rest of the cockloft," I said. Everyone on my crew nodded. We moved to the center of the roof.

"I bet this fire wall has poke-throughs," Gary said. Gary was saying what we all knew to be true, that these older ordinary buildings had fire walls that had been compromised by workers running utilities, like new electrical conduit and cable lines, through them.

I couldn't see the roof or my boots due to the heavy smoke, so I stopped and used the thermal imaging camera.

I showed Ted and Jrob where I wanted them to start cutting. "Right here," I said, pointing with my roof hook through the smoke, letting them look through the TIC.

Everyone went to work. We started in the center of the roof, about three feet from the fire wall. Jrob and Ted used the chainsaws to make louvers over a joist as they moved away from each other. I stayed with Ted, opening the louvers and punching through the ceiling below with my hook as soon as he cut them, and Gary did the same with Jrob. In five minutes or so we had a trench running the entire sixty-foot width of the structure. Fire was just making its way through the fire wall and was beginning to show out the trench as we finished up the final louvers on each end.

I radioed to the IC that we had dropped back and had finished cutting a defensive trench cut on the strong side of the fire wall, and that the fire had breached it. I also asked the IC for two more truck companies and one engine company for a protective hoseline.

"OK Ladder 10," the IC said over the radio. "Ladders 5 and 3 have been assigned to your location, as well as Engine 25. You are now ventilation group supervisor."

"OK Command," I said. "Ladder 10 will be ventilation group supervisor. We also need some fresh air bottles and chainsaws up here, staged at the top of Ladder 9's aerial."

"Let's fall back," I told Ted, Gary, and Jrob. We made our way back over the next fire wall, back towards Ladder 9's aerial ladder, where we met up with the just-arriving crews of Ladders 5 and 3. They slapped us on the backs and pointed to the twenty-foot wall of flame that was now blowing out from the trench cut we had just finished.

"Nice work, boys," Captain Childs of Ladder 3 said as he met up with us. "Looks like you guys had all the fun." He and I discussed cutting another trench on the strong side of the fire wall we were all standing by.

"OK," he said. "We'll do it. We just won't open up the louvers unless it starts coming through this fire wall." He explained to his crew what the plan was, and they started cutting another trench.

"Hey boss," Gary said a few minutes later, as we watched the flames that had been blowing out the windows from floor 4 turn to steam. "We're low on air, and these chainsaws are gummed up with tar."

"OK, let's move back to the aerial. We should be getting some fresh bottles and chainsaws up here any minute," I told him.

As we were making our way back to the aerial, Gary said, "Hey boss, that was some strong work."

"Damn strong," I smiled. "You old guys do pretty good at fires."

Gary shook his head slowly and let out a big laugh. But we both knew what he meant—that I had passed the test.

We were all tired, but it was that good type of tired that firefighters feel when they know that their actions made a difference in the firefight. Soon the cockloft fire was being attacked from below by the engine companies on floor 4. In a few minutes, there was nothing but steam showing from our trench cut.

It wasn't too long before the IC came over the radio, saying, "All companies, fire under control."

Looking back, we had used the construction of the building to our advantage. Once on the roof, we made an inspection cut to determine what the roof was made of, how thick it was, which way the joists were running relative to the walls, and what the fire condition was like underneath us. We had made our louvers "with construction"—that is, we used a 2×10" roof joist as the center of our louvers. By cutting "with construction," we were able to cut larger cuts in the horizontal direction, which allowed us to finish our trench cut faster than it would have taken us if we were working "against construction." We also knew that we could work on this conventional style roof for a long time. Conventional construction utilizes structural components that rely on size for strength, and we knew that the 2×10" roof joists could stand quite a bit of heat and fire.

STRUCTURAL STABILITY

The building's ability to resist collapse is called structural stability. There are several factors that affect a building's ability to resist collapse. These factors are the condition and size of the structural components, the type and condition of the connections which transfer the load from one structural component to the others, the fire load, and the load's heat release rate inside the building on fire.

If a structure is older, or has not been maintained properly, the chances are that its structural components have deteriorated over time. Weather, age, insect infestation, water leaks, previous fires, or a combination of these can weaken or destroy the structural components of a building. When a fire breaks out, these compromised structural members tend to fail more quickly than they would have if they were new and in good shape. Take, for example, a beam that has been infested with termites, which have eaten some of the beam's mass. When this beam is exposed to the heat and flames of a fire, the beam will fail earlier than it would if it had not been damaged and compromised by the termites.

The size of the various structural components of a building influences its ability to resist collapse. The larger the component, the longer it can withstand the heat and flames of a fire before it fails. I talked about this earlier in the discussion about construction styles and the difference between conventional/legacy and lightweight/modern construction. Structural members built using conventional construction rely on size for strength, and are often found in older buildings. Lightweight construction's strength depends on the geometry of its members, not size. In lightweight construction, mass has been replaced by mathematics—that is, dimensional lumber has been replaced by engineered lightweight truss systems. Most modern buildings have lightweight structural members in the form of trusses. The simple fact is that trusses fail much faster in a fire than do comparable conventional structural members, and they have cost many firefighters their lives. Francis Brannigan warned us again and again over his long and distinguished career: "Beware the Truss."

The connections which transfer the load from one structural component to another are vital to the building's stability. The building is only as strong as the connections which hold it together, and if one of these connections fails, then the entire building, or part of it, will collapse. Picture the thin, metal gusset plates, or gang nails, that hold the members of a lightweight truss together. If the wood fibers holding the gusset plate are exposed to enough heat and fire, then the gusset plate can no longer hold the members of the truss together, which can lead to the failure of the entire truss. Another example would be a masonry wall that is holding up an unprotected steel girder (a beam), which, in turn, is holding up the floor joists (also beams). If the steel girder is heated by a structure fire to around 1000°F, then it will elongate and twist, causing the joists to drop and the masonry wall to be pushed out. The simple fact is this: the connections that hold the structural components together are critical to the building's resisting gravity and staying up.

The final factor that affects a building's stability is its fire load and the fire load's heat release rate. *Fire load* is defined as the potential fuel available to the fire. Both the contents and the building itself, if it is combustible—like the lumber in a stick-framed house—make up the fire load. The fire load of a structure represents the total amount of heat in the fuel.

The *heat release rate* of the fuel is the rate, or how fast, the heat is released. The heat release rate of the fire load is very important to firefighters. Why? Because the faster a fire load releases its heat, the hotter and the more severe the fire will be. And the more severe the fire, the faster it reaches flashover and affects the structural components and the connections of the building.

Picture 1000 lbs of 4'×8' OSB sheets (Oriented Strand Board, an engineered sheet of wood formed by adding adhesives and then compressing layers of wood strands in specific orientations), and 1000 lbs of 12"×12" timbers, both housed separately in low sheds that have exposed lightweight truss roofs. Now picture that the OSB and the 12"×12" timbers are on fire inside their respective sheds. The heat release rate of the OSB is many times faster than that of the timbers. The OSB will release its heat much more quickly than the timbers, causing the heat and fire from the plywood to expose the lightweight trusses above faster. Because of the heat release rate, the exposed roof trusses will fail in the shed of the OSB fire long before those in the shed of the 12"×12" timber fire.

All firefighters should be familiar with the basic heat release rates of wood (both dimensional lumber and lightweight TGI joists and truss systems), plastics and synthetics, and flammable liquids. Materials that release their heat quickly can cause severe fire events inside structures.

Heat release rates of building materials and contents have a huge effect on fire growth and fire spread. Rapid fire growth has been responsible for large loss of civilian life in structures over the years. 492 people perished in the Cocoanut Grove nightclub fire in Boston in 1942.[1] A contributing factor was the rapid fire growth from the highly combustible acoustic ceiling tile used in the club. In Los Angeles, 24 occupants of the Dorothy Mae Apartments died in 1982 due to flame spread over plywood in the exit hallway.[2] Rapid flame spread across plywood paneling used for the interior finish at the Elliott Chambers Boarding House in Beverly, Massachusetts, contributed to 15 deaths in 1984.[3] At the First Interstate Bank Fire in Los Angeles in 1988, the floor of origin had wide open spaces packed full of plastic contents—computers and cabinets—which contributed to the extremely rapid fire growth and spread that destroyed floors 12 through 16.[4]

Every firefighter and fire officer must be a career-long student of building construction. As structural firefighters, we work in and around buildings. Structure fires are our job, so we must take the time and energy to know everything we can about them—what materials are used to make them, how they are put together, and how fire will affect them. As Brannigan said, "The building is your enemy. Know your enemy."[5]

Endnotes

1. https://bostonfirehistory.org/the-story-of-the-cocoanut-grove-fire/
2. http://www.usdeadlyevents.com/1982-sep-4-arson-fire-murder-dorothy-mae-apartment-hotel-los-angeles-ca-24-25/
3. http://www.usdeadlyevents.com/1984-july-4-arson-elliott-chambers-rooming-home-fire-beverly-ma-15/
4. https://www.nist.gov/el/interstate-bank-building-fire-los-angeles-1988
5. https://www.fireengineering.com/2018/06/11/202749/building-construction-review/

STRATEGY

The third core principle of structural firefighting is **strategy** (fig. 3–1). The basic strategy of modern structural firefighting is constant—to protect life first, confine the fire, and then extinguish it. This basic strategy can be broken down into three main strategic priorities:

- Life Safety—both civilian and firefighters' lives
- Incident Stabilization—or fire containment and ultimate extinguishment
- Property Protection

Figure 3–1. Fire strategy is the plan which considers life safety, stabilizing the incident, and protecting property.

Think of it this way—the strategy of a firefight is the overall plan. This plan should be based upon accomplishing the three main strategic priorities of life safety, incident stabilization, and property protection.

The strategy for fighting a fire should answer some basic questions. How is this structure fire to be contained, offensively or defensively? Where should the firefighters be deployed? On the outside of the structure? Inside the structure? How should the apparatus that show up at the fire be utilized at the scene? Should the first ladder company to arrive be used for victim rescue or ventilation? Should the equipment and tools be used to extinguish the fire first, or to protect and save any human lives that may be in danger? These questions must be answered by the IC's initial strategy at any working structure fire. And to make sure that we are all on the same page here, the first IC at a structure fire is typically the first-in company officer.

I believe that the initial IC must conduct a thorough size-up of the structure fire *before* they decide upon a strategy. An initial size-up of a structure fire is an assessment of the problems and conditions that are present on arrival. But how is this initial assessment to be done? How should it be conducted?

I would argue that in order to conduct a good initial size-up at a structure fire, the initial company officer must get off the rig and do a 360° walk-around of the building to see what is going on and to look at what is happening on the sides they cannot see, particularly the back side of the building, side C, Charlie. Valuable information, like fire location, victim location, and other hazards and conditions are oftentimes discovered on this walk-around. Sometimes a full 360° walk-around of the structure on fire is not possible due to its size, its position on the block, or the presence of fences, gates, hedges, guard dogs, a ravine or steep slope, or other such barriers or obstacles. In these cases, the officer can ask another unit to make its way to the back side and report what they see. However, if these hindrances do not exist, there is no excuse for the first-in officer not to conduct a walk-around of the structure on fire.

What comprises a thorough size-up? Some fire departments mandate a traditional 13-point size-up using COAL WAS WEALTH (Construction, Occupancy, Apparatus and staffing, Life hazard, Water supply, Auxiliary appliances, Street conditions, Weather, Exposures, Area, Location and extent of the fire, Time, and Height). Other departments use modified versions of this traditional size-up, having the first-in company officer give a size-up over the radio that includes much of the following information: degree of flame and smoke (heavy, moderate, light), location of the fire and/or smoke, if the structure is occupied or if there are obvious victims, height of the structure, construction type, occupancy type (single family, apartment, commercial), size of the structure, exposures, initial strategy and tactics, and command designation.

Whatever your department's standard operating guidelines (SOGs) call for in a size-up, the fact is that you have to take a moment and look—*really look*—at what is going on. And the only way to look is to get off the rig and walk around the structure to see what you and your crew are up against. How else can you do a high-quality assessment?

But what is the first-in officer really looking for?

Lieutenant Mike Ciampo of the FDNY, and author of the always-informative "On Fire" column on the back page of *Fire Engineering* magazine, proposes a quick, easy way of conducting a rapid initial size-up. He argues that the first-in officer should concentrate on three key priorities when making an initial size-up: life, fire, and operations.

Ciampo points out that since life is our first priority at any structure fire, then the first-in company officer must account for life in their size-up. Are there any victims present at any windows? Are there 911 reports of people trapped? Are there cars in the driveway? Is a family member or neighbor out on the street yelling that the occupant is inside? If any of these cues or conditions exist and fire conditions permit, Ciampo asserts that the initial strategy must include a safe and aggressive interior search.

Fire is the second priority in a rapid initial size-up. What kind of fire is it—contents or structure? Where is it located? What is the building type and construction? What is the path of fire travel? Where is it going? What is the fire exposing? Is it a big fire that needs a 2½" hoseline? Or is it a small fire that can be handled with a 1¾" line? Ciampo reminds us that we also must read the smoke, like Dave Dodson has been preaching for years. What is the smoke from the fire telling you? Is it exiting a vent point with velocity and turbulence? What is the volume of smoke? What color is it? Is it brown and thick, telling you that the fire is most likely burning the wooden structural members? Or is it black and thick, telling you that the plastic and synthetic contents are burning?

Operations is the third priority of this quick size-up. Ciampo points out that having standard operating guidelines (SOGs) for specific structures, with predetermined assignments and procedures, can help the first-in company officer in their initial size-up decisions. An example of this would be that first-in officers are to deploy a 1¾" line for residential structures and deploy a 2½" line at commercials. Procedure-based assignments work for many departments around the country, and they can help make the job of sizing up a structure fire a bit easier.

The initial strategy at a structure fire must be based upon the three main strategic priorities, which are life safety, incident stabilization, and property protection. The IC must also base their initial strategy on a rapid and ongoing risk assessment to their crew.

Life Safety

Protecting lives is the primary mission of any fire department, and it is the first priority of firefighting strategy. To protect and save human lives is the job we all signed on to do. Saving lives is our highest calling, our sworn duty, and the main reason the public we serve respects, honors, and reveres us so much. The civilians we serve recognize that when they are in great danger, firefighters will put themselves in harm's way to save them. And sometimes, sadly, firefighters end up losing their lives trying to save civilians in structure fires. However, we must keep in mind that firefighter lives are just as valuable as civilian lives, and that firefighter lives are not to be risked when there is little or nothing to gain. The first responsibility of all fire officers should be the safety of the personnel under their direct supervision

Firefighters should expect to take reasonable risks to their lives at a structure fire to save the lives of civilians who are truly in peril. In other words, firefighters will risk a lot—their lives—to save a lot—the lives of the public they serve. However, a rapid risk assessment must be made by the first-in officer during their size-up. If the risks to the firefighters are taken into account and are worth the benefit or reward, such as saving a human life, then the officer is justified in sending firefighters into the building on fire.

Suppose an engine company arrives at a structure fire with flames visible from most of the first floor of the house and heavy smoke showing from a second floor window on side D, Delta. A mother is outside the house frantically screaming that her child is up in the room that has smoke showing from it. The degree of risk is high—a well-involved fire on the interior of a structure. But the benefit is also very high—saving the life of the child who is up on floor 2 exposed to the smoke and heat of the fire. In this case, the reward is worth the risk to the firefighters. There is a known victim in a known location. The officer and firefighters on that engine company will do everything they can to save that child on floor 2. They will throw a ladder to the floor 2 window on side D and pull a protective line. One or two firefighters will climb up the ladder, clear the window, enter the room, and attempt to rescue the missing child.

Now suppose the same engine company arrives at the same house, but the fire conditions are drastically different. The entire house is on fire, with flames visible from every window and doorway. Flames are through the roof in the back portion of the house—the fire has had a long head-start. The mother is outside the house frantically screaming that her child is up in a room on the D, Delta, side of the house. Heavy flames are venting from the top to the bottom of this window. The risk/reward equation is much different now. The risk of sending firefighters on an aggressive interior attack or interior search of the floor 2 bedroom is very high. And the reward is extremely low. The chance of the child being able to survive the fire conditions in the bedroom is virtually impossible. In this case, the officer of the engine company must first think of the safety of his personnel and not put them at great risk. Unfortunately, the survivability profile for a fully involved room that has been free-burning for a long time is close to zero. An exterior, transitional attack on the fully involved house would be a reasonable initial strategy here, keeping the firefighters in a low-risk environment until the fire is sufficiently knocked down and it is a lower risk for them to enter the house and fully extinguish the fire (fig. 3–2).

What Is Acceptability of Risk?

I remember arriving on the second engine company at a fully involved house fire in the central district of the city. I was working a shift on Engine 30, an engine company that sees quite a bit of fire. The engine company that arrived first took command and began to attack the fire from the exterior with a 1¾" line. The officer on this engine company was newly promoted and lacked experience. He failed to do a good size-up of the fire and neglected to report his strategy. Was this a defensive fire attack? An offensive attack? A transitional attack? No one listening to his radio report knew what he was doing or thinking.

When I arrived on Engine 30, I saw a two-story, wood-framed, abandoned house slated for demolition that had heavy flames coming from most of the windows, exposing a house on the B, Bravo, side. I told my crew to pull a 2½" exterior

Figure 3–2. An exterior, transitional attack in a multi-dwelling building.

attack line and a 1¾" exposure line to protect the house on side Bravo. One of my crew took the charged 1¾" line and began to wet down the house on the Bravo side, while the other members took the 2½" line and began to put water on the house fire from the front yard through a large picture window—an exterior attack position.

We were darkening down the fire on first floor with our 2½" line when, without any warning or reason, the first engine company's crew on the 1¾" line began to enter the house from the front door. I ran over to the first-in officer, who was the IC, and asked him why his crew was going into this house, which I considered a loser—again, it was slated for demolition with significant fire involvement.

"I don't know," he told me, with a blank look on his face. "What should I do?" It was obvious that he was in way over his head.

Just then I heard the crew of this engine company report over the radio that they were making headway on floor 1 and were moving up to floor 2. The first ladder company arrived, Ladder 3, along with the third engine company, Engine 2. I tried to call the first-in engine company over the radio, but they didn't answer.

My crew members on the 2½" line had shut it down the moment they saw the first-in engine company entering the house. The life safety issue was now the firefighters from the first-in engine company—they were inside a loser building with heavy fire involvement. And they were not responding over the radio.

I had no choice but to risk my firefighters' lives in order to back up the first-in engine company on the interior. I told my crew to abandon the 2½" line and take the 1¾" line into the structure. Ladder 3 went to the roof to open up several vent holes over the well-involved attic, and Engine 2 also came into the house with a line. I remember being forced down to the floor by high heat up on the second story at one point, asking myself what the heck we were doing in this house. It took all the companies at the scene a long time, but we finally extinguished the fire. It wasn't until later that we discovered that the house was being used by homeless squatters who were using heroin. We found dozens of needles on the floor and drug paraphernalia in several of the bedrooms during overhaul—another reason not to have entered this vacant structure.

The end result of this fire was that many firefighters' lives were placed at high risk for a vacant building that was demolished one month after the fire. The first-in officer had failed to decide upon a strategy and did not control his crew's actions. To make matters worse, the battalion chief who arrived at this fire did nothing to change the high risk, offensive strategy that the first-in engine company had forced everyone to follow. What should have been a no-brainer defensive fire turned into a high-risk fire that pointed out a failure of risk assessment, initial strategy, and command and control. In the end, we were lucky that no one got hurt. As Deputy Chief Vincent Dunn, who retired from the FDNY after a forty-two-year career, has stressed over and over again in his lectures and numerous books, "No building is worth the life of a firefighter."

Fire Containment

Once the life safety priority has been addressed and handled, the second priority of firefighting strategy is fire containment, which will ultimately result in extinguishment. An initial fireground strategy must keep the fire from getting worse than it already is. The goal is to control it and to keep it from spreading. The IC must make sure that they properly evaluate the fire and keep it from escalating so that more lives are not threatened. This may mean writing off the initial structure as lost and protecting the exposures that are adjacent to the structure on fire. It may mean letting an entire floor of a high rise burn all its fuel while protecting the floors above it or the stairwells for occupant egress. Or it may mean writing off a row or block of houses and protecting the next row or block (fig. 3–3).

After accounting for life safety, keep the structure fire from getting worse than it already is. Contain it, if this can be accomplished with moderate risk to on-scene personnel. In order to keep a structure fire from getting worse, the first companies on scene might be forced to confine or contain the fire until sufficient resources arrive at the scene to extinguish it.

Picture a fully involved fire in the basement of a large, occupied apartment house during the early morning hours. The fire started in the laundry room, has extended out into the basement hallway, and has now involved the storage rooms and made its way up the stairwell to floor 1. Heavy black smoke is filling the floor 1 hallway because someone propped open the stairwell door to the hallway. This stairwell serves the entire building, from the basement to floor 5, and is the only means of egress for the tenants of this building.

The first-in engine company has a choice to make: take a charged line to the basement entrance or window and fight the well-involved fire, or take a charged line inside the apartment house to protect floor 1 and keep the fire from making its way further up the stairwell.

The right choice would be to take the charged line inside the apartment house to protect floor 1 and the stairwell. Why? Because by doing this, the first-in engine company is containing the fire to the basement and not letting it get any worse than it already is, cutting off its vertical extension. The second-in engine company can attack the basement fire from the basement entrance or a basement window, while the third-in engine company can provide a backup line for them. Though the actions of the first-in engine crew seem secondary and not important, they are setting up the entire fire operation for success by keeping the fire contained to the basement area.

Sometimes it is hard to watch a room or a floor or a structure burn without putting water on it. It takes discipline, training, and a view of the total picture or strategy. But if you show up with a single engine company to a burning structure that is already lost, or will be soon, and a neighboring structure is about to catch

Figure 3–3. Ladder 9 uses a Ladder Pipe at an apartment fire.

fire, your choice must be to contain the original fire and keep it from getting worse by saving the exposed building.

I was on Engine 31 in the north end of Seattle one night when we were dispatched to a working fire in another engine company's district. We were the second-due engine, and as we were driving to the address, we listened to the initial report from the first-in engine. The officer reported flames visible from all sides of a two-story, wood-framed house. He then said that the engine was on a hydrant, and that they were laying a 1¾" preconnect to attack the fire. My officer, Capt. Smith, commented that there was no information about any exposures, and he wondered aloud if the fire was in danger of spreading to any of the adjacent homes on the street (fig. 3–4).

Figure 3–4. Firefighters on a patio keep a fire from spreading to other residences.

When we arrived, I saw a fully involved house fire, with a hose team from the first engine company using a 1¾" line from the front yard and having no effect whatsoever on the body of fire.

Before Engine 31 could even stop, a man came running up to us, waving his arms.

"Help! My house is gonna catch fire," he yelled, pointing to the house to the left, side B, Bravo, of the original house on fire. The cedar siding on the side of his house closest to the fire was off-gassing from the radiant heat and was about to light off.

Captain Smith surveyed the scene quickly. "Steve, Joey, let's take two 1¾" lines off the first Engine, one to the Bravo side of the house, and one on the Delta. We're going to have three houses on fire in a minute if we don't get some water on them."

Then Captain Smith told our driver, Bubba, to get a second supply for the first engine, and then pull a 2½" line off the first Engine and start knocking down the fire.

Captain Smith took command of the fire and began to get everything rolling the way he wanted. Joey and I each took a 1¾" line and began operating them on the exposure houses. I wetted down the entire side of the house that belonged to the man who had asked us to help him. The house stopped off-gassing immediately and was no longer in danger of catching fire. Joey did the same to the other exposure. Bubba had pulled a 2½" line, and with the help of Engine 17, was beginning to make headway on the house fire. Soon there were multiple 2½" lines working. It wasn't too long before the house fire was knocked down and under control.

The officer on the first-in engine made several key mistakes. First, he forgot that after life safety, the next priority is to keep the fire contained, to protect the exposures. He should have pulled lines and protected the houses on the B and D sides of the house fire. His next mistake was to pull the wrong sized hoseline to attack the fire. He should have pulled a 2½" line or used the deck gun. A 1¾" line for a fully involved two-story house does nothing to make the situation better. A big fire requires big water.

Fire containment certainly isn't always sexy, and it might seem boring, but after life safety, it is the most important incident priority. Don't let the fire get bigger. It seems so simple, but I am amazed at how often this priority is overlooked.

I recall watching another big city fire department ignore fire containment and concentrate on the fire. The fire started in a car under a carport in the early morning hours and spread quickly to the other cars. Soon there were six fully involved cars under the carport, and the fire was exposing the apartment house across the driveway.

The first engine company arrived, pulled a 1¾" line, and began attacking the car fires. Unfortunately, the 1¾" line was too small for such a large amount of fire—six cars fully involved under a wood framed carport. Their initial actions did nothing to knock the fire down. Instead, the fire continued to expose the roof eaves of the occupied two-story apartment house. Soon the attic of the apartment house was on fire, as well as the carport and cars. In the end, the entire roof was lost, and the tenants of the apartment house had to find other living arrangements until the apartment was repaired.

Property Protection

After life safety and fire containment, the next strategic priority is property protection—saving property. Certainly, saving property from fire is part of our job. By saving property, firefighters keep civilians in their homes and jobs. However, firefighters' lives should never be put at extreme risk to simply save property. Some might argue this point, saying that saving property is just as important as saving lives. I disagree. This type of thinking ignores the concepts of risk assessment and risk management. Evaluating the risk to firefighters' lives must be part of the first-in officer's initial and ongoing size-up, and has to be part of the critical decision making process. First-in officers at a structure fire must do a proper risk assessment and only commit their firefighters to saving property from fire if the risk to their firefighters is low. As the late Chief Alan Brunacini of the Phoenix

Fire Dept. stated, "We will risk our lives a lot, in a highly calculated and controlled manner, to protect a savable human life. We will risk our lives a little, in a highly calculated and controlled manner, to protect savable property. We will not risk our lives at all to protect lives or property that is already lost." *Savable* human lives and *savable* property are the key concepts of Bruno's rules of engagement. If the survivability profile of the fire tells us that there are no savable lives in the structure on fire, and if property is not savable, then no firefighters should be put at any risk. Chief (ret.) John Norman also makes this same point in his discussion about the General Principles of Firefighting in his exceptional, classic book, *Fire Officer's Handbook of Tactics.* In his 5th Concept, Norman states, "When there is no threat to occupants, the lives of firefighters shouldn't be unduly endangered." In other words, firefighters must not rush headlong into an aggressive interior attack if all the occupants are out of the structure and accounted for, or if the building is a known vacant structure or is known to be unoccupied.

Chief (ret.) John Mittendorf brought up the topic of property conservation one night when we were having a wide-ranging discussion on the art and science of structural firefighting. We were talking about fires in commercial occupancies when he mentioned that many of the large big box stores do not carry any fire insurance on their buildings. Their risk management people have determined that it is cheaper to build a new structure after a fire than it is to pay the fire insurance. Mittendorf made the argument that if these big box stores don't care about their property—they are not insuring them—then why should we ever put firefighters inside them if they are on fire and we can confirm that there are no people inside. "These are truly disposable buildings," Mittendorf said.

This idea of disposable buildings is something the fire service needs to understand and consider when fire companies respond to a fire in one of these structures. Take, for example, a fire in a fast food restaurant. If there is a fire of any significance in this fast food restaurant, the structure will be quickly demolished and rebuilt—we know this to be true. And if we know that this structure is disposable, then why risk any of our lives trying to save it? Firefighters' lives have been lost at fast food restaurants, as happened on February 14, 2000, in Texas, where two firefighters perished.[1]

The idea of disposable buildings seems odd and foreign to many firefighters who have been around for twenty or more years. In my city, Seattle, we have many ordinary constructed buildings well over one hundred years of age. Most of these have been remodeled and upgraded over the years, but they are still around, and a good percentage of them have had fires inside them over the years. Some of these structures have had multiple fires inside them over their lifespans. The idea that someone would design and build a structure that could be disposed of after a fire of any significance is a new concept to many firefighters and fire officers, but it is something that the fire service must understand and come to terms with.

These disposable buildings are, for the most part, constructed using lightweight components, which we know are not firefighter-friendly. As I pointed out in Chapter 2, the mass of structural components have been replaced by mathematics—dimensional lumber has been replaced by engineered lightweight truss systems.

As much as firefighters enjoy fighting and extinguishing fires in buildings, the risks have simply become too high to continue charging into buildings that don't have occupants inside them or buildings with fire environments too severe for life to survive. At every structure fire, fire officers must keep firm control over the actions of their firefighters, keeping them from risking their lives needlessly for property that is already lost, property constructed of lightweight components weakened by fire, or property that will be demolished and replaced soon after the fire is out. We owe this to the memories of all of our brother and sister firefighters who have lost their lives in vacant, unoccupied, lightweight, or disposable structures.

This brings us back to the fact that the first-in officer, the initial IC, must do a risk assessment as part of their size-up at every structure fire. And this risk assessment must inform the strategy of the structural firefight. Without a realistic risk assessment, the strategy of the firefight is incomplete and may be putting the lives of firefighters in peril for no good reason.

Risk/benefit analysis, risk vs. reward, risk assessment, risk analysis . . . all these mean essentially the same thing and ask this one basic question: what is to be gained by putting our firefighters at risk at this particular structure fire?

I recall watching video of firefighters from another city attacking a church fire. The church was a relatively new building and was unoccupied—it was the middle of the week and the parking lot was empty. The fire was well-involved in the attic space above the sanctuary, and it was obvious the roof was held up by trusses. The first two engine companies to arrive went inside the structure with 1¾" handlines in an attempt to extinguish the fire. Unfortunately, the steeple soon collapsed onto the trusses above the firefighters who were working inside to extinguish the large volume of fire attacking the structural components. The entire sanctuary roof came down, on fire, on the firefighters inside. One firefighter lost his life, unable to escape the inferno. The official cause of death was asphyxiation—he ran out of air, pinned down by the collapsed trusses.

Was a risk assessment performed in the initial size-up? What was to be gained by putting firefighters at extreme risk in an offensive, interior firefighting strategy?

Sadly, nothing was to be gained. The structure was unoccupied. The fire was well advanced in the attic space, severely weakening the trusses that held up the roof. The structure was going to burn to the ground anyway, no matter what strategy was chosen. And yet, firefighters were allowed to go inside this loser building, telling me that an aggressive, offensive strategy was either decided upon by the first-in officer, or that it was the default strategy of this particular fire department. And the end result was that a firefighter lost his life and the building burned down.

When will we finally learn the hard lesson that Vincent Dunn, John Norman, and Alan Brunacini having been telling us for years, that no building is worth a firefighter's life?

Endnote

1. https://www.cdc.gov/niosh/fire/reports/face200013.html

TACTICS

The fourth core principle of structural firefighting is **tactics**. Tactics are the actual hands-on operations that must take place in the correct sequence to successfully accomplish the overall strategy of the firefight (fig. 4–1).

Think of tactics this way, using a sports analogy—tactics are the operational plays that must happen in order to carry out the strategic game plan.

Tactics are the physical actions engine and ladder companies do—pulling hoselines into structures, getting a water supply from a hydrant, cutting holes in the roof for ventilation, throwing ladders to windows for rescue, searching an occupied dwelling for victims with a Thermal Imaging Camera, forcing entry into a secured building with the irons or the rescue saw, using a positive pressure fan

Figure 4–1. Tactics are the firefighting methods used to achieve the strategic goals of the incident commander.

to keep smoke out of an attack stairwell, protecting exposures with a deck gun or monitor, using a ladder pipe to put a lot of water on a large fire, etc.

Unlike the strategy and the strategic priorities of structural firefighting, which are generally constant and unchanging, the tactics of modern firefighting often do change, oftentimes due to advancements in technology. The tactic of searching an occupied structure full of blinding smoke has been revolutionized by the Thermal Imaging Camera (TIC). The TIC allows firefighters conducting a search to see through the smoke. A large, smoke-filled room, which used to take minutes to physically search by crawling around the room and sweeping with a hand tool can now be searched in seconds using a TIC.

Think of all the advancements in technology that have changed our structural firefighting tactics over the years. Positive pressure fans have forever transformed the way we ventilate certain types of structure fires. Today's SCBA and bunking gear allow firefighters to get inside structures quickly with a hoseline and attack the seat of the fire. Decades ago, an aggressive interior attack of a structure fire was nothing less than a heroic feat, done with minimal thermal protection and without a reliable supply of fresh air. The bucket brigades and the steam-powered pumpers of the past have given way to impeller-driven fire pumps on today's modern fire engines, allowing the driver—the chauffeur—to deliver water under pressure to multiple handlines and appliances. Chainsaws and power saws have replaced axes and allow ladder companies to quickly ventilate flat, arched, and pitched roofs. Standpipes in multistory buildings have replaced the labor intensive task of stretching hand-laid supply lines up stairwells. Today's standpipes, along with fire pumps, are fixed, reliable water delivery systems that save both time and labor, and have changed the tactics of fighting fires in tall buildings.

This brief study is not meant to be a book on structural firefighting tactics—there are far better texts on engine and ladder company tactics written by the likes of Norman, Dunn, Brennan, Mittendorf, Brunacini, Salka, Avillo, and many others. However, I would like to spend a little time discussing *basic* engine and ladder company tactics at structure fires.

Engine Company Tactics

The most important tactics of the first-arriving engine companies at a fire in a building are to get the proper size and amount of hose to the fire location, secure a water supply quickly, and get water on the fire as rapidly as possible.

But let's break down these basic engine company tactics and discuss them on by one.

Size-up and Operational Risk Assessment

The officer on the first-arriving, or first-due, engine company must conduct a rapid, initial size-up that includes a risk assessment. Remember, an initial size-up of a structure fire is an assessment of the problems and conditions that are present on arrival (see Chapter 3), and the only way to do this is to have the first officer get off the rig and look—really look—at what is going on.

Apparatus Positioning: The Front of the Building Belongs to the Truck

Apparatus positioning is everything!

The engine companies arriving at a fire in a structure must leave the front of the building open for the incoming ladder company. The ladder company needs this position to accomplish the necessary tasks to support the engine companies: forcible entry, laddering the building for entry, removal or rescue (both civilian and firefighter), ventilation operations, search, getting to the rear—or back, side Charlie—of the fire building, checking for fire extension, and overhauling the fire when it is over. Think of the ladder company as a big tool box. This tool box must be located in front of the building so that the "tools" are close by and can be used in a timely fashion. You can't use an aerial ladder if the ladder company can't get close to the fire building.

In order to give the front of the fire building to the truck, the first-arriving engine must either stop short or pull past it. Stopping short or pulling past the fire building gets the engine out of the way so the truck company can get the position it needs: the front of the building. In terms of securing a water supply, this means that the first-arriving engine company must do one of three things. The engine company must make the stretch from the hose bed and bring the water supply from either a hydrant or another engine on the hydrant to the attack engine's pump via supply hose. This can be done by laying forward, or by overhauling the supply line to the hydrant or supply engine. (Here in Seattle, our SOGs dictate that if the attack engine is over one hundred feet from the hydrant, then the next engine will park on the hydrant and supply the attack engine). The third option is to have the engine company drop a hoseline (or hoselines) from the engine near the fire and then lay reverse to the hydrant to secure a hydrant supply.

When I finished recruit school and got assigned to Engine 31, Captain Smith taught me that the first-arriving engine company should always lay reverse to the hydrant. I remember arriving first to many fires where we dropped our manifold—which was attached to our 4" supply line—and our hoselines, and then had the driver lay out to the hydrant. This was our standard hydrant lay, unless there was a hydrant that the driver could easily reach and overhaul by himself, in which case we would pull preconnected lines off the rig. Captain Smith was always preaching that was easier for the pump on the engine to push water from the hydrant to our hoselines rather than to pull water, and that the front of the building belonged to the ladder company, which a reverse supply always ensured.

Today we know much more about fire behavior than we did thirty years ago. Reverse lays usually take time—time that could be better spent putting tank water on the fire with a preconnected line and cooling it down as quickly as possible. This tactic has replaced reverse lays in many parts of the country. However, engine companies from the Detroit Fire Department, companies that see more structure fires than almost any other city in America, lay reverse out to the hydrant at almost every structure fire they fight.

Confirmed Victims in a Known Location

Arriving first at a fire in a structure, an engine company officer must account for any confirmed victims who are truly in peril and whose location in the structure is known. These are victims you can see with heavy smoke or fire pushing behind them at windows or balconies, or victims you cannot see whose location is known by someone standing outside—like the mother screaming that her child is up in the second story bedroom as she points to a window with smoke showing from it.

Now this confirmed victim may need to be rescued from a second story room that has heavy smoke under pressure showing from it—again, picture the mother pointing to the window. If the officer on first-arriving engine company showed up knowing that no other rigs were going to get to the fire location anytime soon, they would quickly throw a ladder to the window and rescue this victim in immediate danger from the fire, while also pulling a protective hoseline. However, if the first-arriving engine company had a ladder company showing up right behind them, the engine company officer would let the ladder company throw the ladder and rescue the child, instructing the members of the engine company to stretch a protective hoseline to the window to protect the victim and ladder company—remember that there is heavy smoke coming from this window. After the rescue of the known victim on floor 2, the engine company could easily drop the protective line, and then pull another hoseline to the front door and move into the house to put out the fire.

Picture another scenario where a man is leaning out from a second story apartment window with very light smoke showing from it, and the smoke is not under pressure. Flames are blowing out from a second story window, one apartment over from where the man is screaming for help. People exiting the apartment are telling you that there is smoke in the center hallway. In this instance, the first-arriving engine company should quickly place a handline into operation and get water on the fire in the floor 2 apartment. The man screaming for help is not in any serious danger, at least not yet. By putting out the fire, the engine company is removing the danger from the screaming man, and, in effect, saving him.

They are also getting rid of the problem—the fire—as well as saving the structure and any other lives that cannot be seen.

Locate the Fire

When there are no confirmed victims in known locations, the first-arriving company must locate the fire before they stretch any hose. Why? *Because if you don't know where the fire really is, you could well be stretching the line to the wrong location.* This happens with undisciplined and untrained engine companies.

I once watched a first-arriving engine company pull a line up to the fourth floor of an apartment house fire because they saw heavy smoke coming from a floor 4 window. What they missed was the fire raging in a second floor apartment. The door to the fire apartment on floor 2 was left open, and the smoke was traveling up the far stairwell and exiting the building through a floor 4 apartment, which had its door to the hallway propped open because it was being painted by the apartment manager. The firefighters from the first-arriving engine had taken the near stairwell, and the doors to the floors were all closed and therefore smoke-free. The officer on the second engine took their time to locate the fire first—again, by conducting a proper size-up—and had their company pull a line to floor 2 and extinguish the fire, much to the disappointment and embarrassment of the first-arriving engine company, who never could locate the fire on floor 4.

Oftentimes, particularly in larger structures such as apartment houses, high-rise buildings, or large commercial structures, the job of locating the fire is carried out by the first ladder company. If your fire department or fire district does not have a dedicated ladder company, locating the fire must still be accomplished by someone—typically an engine company.

At many fires, particularly single-family dwelling fires, the location of the fire can be obvious, but certainly not all the time! If the fire has self-vented, its location, or at least where it is venting, is visible. Oftentimes, the first-arriving engine company officer is confronted with heavy smoke issuing from multiple locations of the structure: windows, doors, soffits, attic vents. Recall from Chapter 1 that every fire of any significance inside a structure is likely under-ventilated—what we call a ventilation-limited fire. In these situations, knowing the location of the fire is critical to the safety of your crew. Reading the smoke, as Dave Dodson has been preaching for years, is invaluable in these circumstances. If you know how to read the smoke, you can get a pretty good idea of where the fire is in the structure and what type of material might be burning. But this takes practice and dedication—in other words, ongoing training in the art of reading smoke.

A note of caution here: I believe that it is wise to rule out a basement fire at every structure fire. Why? Because many firefighters have lost their lives when they heedlessly entered the first floor of the fire building and then fell through the weakened floor into the inferno of a basement fire. This is why a good, initial size-up at a structure fire is so important. Again, I am a big believer in a 360° walk-around of the structure on fire—it tells you so many things.

I remember arriving at a house fire early one morning. I was the officer on the first-arriving ladder company, working a trade on another shift. The crew working for me that night was one of the best in the city. My driver, Rod, a veteran truckman, noted that the first-arriving engine company had parked right in front of the house as we were pulling up.

"Well, Lieutenant," he said, with sarcasm in his voice, "I guess we don't get position." He nosed in as best he could, but we were not going to be using the aerial tonight.

The tillerman, Jones, muttered over the headset, "You know . . . I hate those engine guys."

When I got off the rig, I saw a two-story house with heavy smoke coming from both floors. An uncharged 1¾" preconnect had been laid from the engine to the front door. The officer of the engine was in the front yard, giving his size-up report over the radio. The driver was at the pump panel, and the other two members of the crew were standing on the front porch with the uncharged hoseline, looking at the front door.

While Rod and Jonzie were getting their SCBAs on, I told Alan, my partner sitting in position #3, to follow me while I walked around the house and conducted my own size-up. But before I could even get started, one of the guys from the engine company standing on the front porch ran over and grabbed me by the shoulder.

"Hey Loot, we can't get in the front door," he said. They needed forcible entry. I recognized the firefighter. He only had a couple of years in the department, and his partner had just gotten off probation. They didn't have a lot of experience in the art of firefighting.

Alan had the irons in his hand, so he and I went over to the porch. I watched as Alan and the youngest member from the engine crew forced the door. As they were working on the door, the hoseline was charged. Soon the front door was open. Heavy black smoke immediately began to make its way out of the front door.

"Command from Ladder 10," I said over the radio. "Can you confirm that we don't have a basement fire?" I wanted to make sure that the engine officer had done a 360° walk around, since I hadn't gotten a chance to do one. I didn't want to be crawling over a basement fire that had been burning for who knows how long.

The initial engine officer, who was still in the front yard and still in command, came over the radio and announced that this wasn't a basement fire.

"OK," I said to the engine company, now holding the charged 1¾" line, and to Alan, "Everyone follows me." With that, I masked up, turned on the Thermal Imaging Camera (TIC), and began crawling into the house. I had a Halligan in my left hand, and the TIC was hanging off my SCBA assembly so I could grab it and use it easily.

It was hot and dark inside. I moved in to the right and scanned the ceiling with the TIC. I wanted to see how hot it was above us. The TIC said it was over 800°F.

"Let's hit that ceiling and cool it down," I told the engine crew. "It's way too hot."

The firefighter on the hoseline opened up the nozzle and penciled the ceiling ahead of us with short bursts of water, which served to cool the overhead and not disrupt the thermal balance significantly. He kept at it for a while, until water began to bounce off the ceiling and back onto the floor. This meant that the ceiling was cooling down, since the water wasn't being instantly converted to steam.

I picked up the TIC again and began crawling to the right, searching for the seat of the fire. If the fire wasn't in the basement, then it had to be on floor 1, most likely back in the kitchen.

As I scanned the floor level ahead of me with the TIC, bright red filled the screen. Fire was issuing from floor level. I sounded the floor with the Halligan as I crawled toward the fire I had seen on the TIC. I looked through the TIC again and saw that we had heavy fire coming from a hole in the floor, about two feet wide. The fire in the basement had burned through the floor, and we were directly over it. I told the Alan and the engine crew what was happening, and kept them back from the hole. I crawled back to their location as quickly as I could.

"Urgent—All crews on the fireground from Ladder 10," I said over the radio, "Be advised that this is a basement fire, with a large hole burned through floor 1 off to the right of the front door, about fifteen feet in."

I instructed the hose team to direct a straight stream into the hole in the floor and get as much water on the fire as possible from our location, about eight feet from the hole. The fire was quickly darkened down, and we began to back out. Engine 25 soon came over the radio, saying that they had a line into the basement from the rear of the structure and had water on the well-involved basement fire. I knew that Lt. White, along with Rich and Cory—the seasoned crew of Engine 25—would have the fire under control in a short time. Ladder 10 team B—Rod and Jones—came over the radio and told me that they were taking out the basement windows opposite Engine 25's hose stream to provide ventilation and were throwing ladders to floor 2.

When we were back near the front door, I had everyone check their air. We all had over half a bottle left, so I said that we were going upstairs to conduct a search. We quickly made our way upstairs. I directed the engine company to stay in the hallway while Alan and I searched the bedrooms. We conducted a quick, thorough search using the TIC. We didn't find anyone, which I told command over the radio. Soon we were all down the stairs and back outside the house.

After the fire was tapped, I went back into the house. Most of the floor joists holding up the first floor living room, where I had seen the fire burning through the floor, had been completely compromised by the basement fire. The joists were burned through. In fact, the plywood subfloor was sagging badly toward the burned-through hole. We had been very lucky we didn't fall into the basement fire below us.

Later, behind one of the rigs, I had a private meeting with the initial engine officer—the guy who initially had command and told me that there was no basement fire—and my battalion chief. I let the engine officer know that his misinformation, his lack of a size-up, could have easily killed four guys that night. I was hot. Truly, this was as close to buying it at a fire as I ever wanted to come. Luckily, the battalion chief calmed me down, and we all talked about what we learned and what went right. It was a teachable moment for the initial engine officer and myself, and I think that we all walked away learning some valuable lessons.

Looking back on it, I must take some of the blame myself. I got distracted from doing my own 360° walk around and size-up when the member from the initial engine company told me that they couldn't get in the front door. After helping them force the front door, I should have finished doing my size-up. This goes back to my main point: when you arrive first at a structure fire, you should always rule out a basement fire.

Get a Hoseline into Position

After accounting for confirmed victims in known locations and locating the seat of the fire, the initial engine company officer must then get a hoseline into position. Tom Brennan, perhaps the greatest fire tactician of our time, has stated that "the proper size and amount of hose, the proper hydrant hookup for the size-up, and quick and rapidly established water supply probably have saved more lives at structure fires than any other tactic."

Here Brennan is saying that if the first-arriving engine company can get the appropriate sized hoseline into operation at the correct location with a reliable water supply, then they will do more to eliminate the life hazard at a structure fire than any other single tactic. And I completely agree with this. Remove the fire, and you remove the threat to life and property.

Remember, unless there are victims in genuine peril and your engine company arrives first to the fire with no other rigs close behind, your best bet is to get the right hoseline at the right location into operation *as quickly as possible*.

In light of the recent UL and NIST studies led by Steve Kerber and Dan Madrzykowski, hitting the fire from an outside position of safety for twenty seconds or so in order to "reset" the fire will lower the interior temperatures dramatically. This *transitional attack* from the exterior makes the ensuing aggressive interior attack that much safer and more effective (fig. 4–2).

Figure 4–2. Transitional attack.

And because of the UL and NIST studies, we now know that the sooner water is applied, the quicker the temperatures decrease inside the structure for those people who may be trapped inside. We also know that applying water from outside will not push fire or endanger trapped occupants. A fast transitional fire attack—applying water quickly from a safe (outside) location—has proven to be the best option for quickly lowering the heat inside the structure and making the situation better for both civilians inside the building on fire and firefighters going inside to find them and extinguish the fire.

The International Society of Fire Service Instructors (ISFSI) have incorporated the lessons learned from the most recent UL and NIST studies of fire behavior into a modern tactical operational plan for the first-arriving engine company. This plan is distilled into five tactical steps that should be followed in order:

- **S**ize-up
- **L**ocate the fire
- **I**dentify and control the flow path
- **C**ool the fire from a safe location
- **E**xtinguish the fire
- **R**escue and **S**alvage, which are "actions of opportunity" that can be initiated any place along the timeline.

The SLICE-RS acronym is easy to remember and takes into account everything we have learned about modern fire behavior in structure fires. If you have not viewed the excellent SLICE-RS video hosted by Eddie Buchanan, you need to do so—NOW! Search YouTube for the **SLICE-RS video from ISFSI**.

Cooling the fire from a safe location does not always have to be from the outside of the structure. Imagine a fire on the tenth floor of an apartment building, where the reach of exterior hose streams is not possible. Firefighters could safely cool this fire from the hallway, directing water into the apartment from the door to the uninvolved hallway. Once they have knocked the fire down and cooled it off from the hallway, then they could enter the apartment and extinguish any remaining fire.

I am personally committed to the ISFSI's SLICE-RS plan and have used it with success as the first-arriving engine officer at a house fire. This modern fire attack plan works and will help reduce the risk to your crew.

The Backup Hoseline

Before a backup hoseline gets placed into operation, the first hoseline should be in position and putting water on the fire. This idea of the second-arriving engine company trying to "steal" the fire from the first-arriving engine should not be tolerated on the fireground. If the first line is having trouble getting into operation, the second-arriving engine company *must* help the first engine company get that first line into operation. *Nothing could be more critical.* Of course, this is assuming that the crew on the first hoseline has located the fire and is heading in the correct direction. If they pulled a line without knowing the location of the fire and they are way out of position, like being two floors above the fire in a large apartment house (see above), then that is their mistake—it would take longer to reposition their handline than to pull another one to the correct fire location. However, if they are hung up on a stairway, or are having trouble getting the line into position due to obstructions or poor hose management or because the driver charged the line too soon, then you must help them, since time is of the essence.

There is an old saying in the fire service that goes, "As goes the first line, so goes the fire." In other words, if the first line doesn't get water on the fire, for whatever reason, then things can get ugly fast. If the crew on the first hoseline needs help getting it into operation, help them, and *then* get your line, the backup line, into operation. Do not allow false pride or one-upmanship or bragging rights to rule your actions—because that is exactly what you are doing if you arrive on the second engine company and fail to give the crew on the first hoseline a helping hand. The citizens you are sworn to serve deserve better.

Choose the Correct Line for the Job

Just as you would never take a knife to a gunfight, engine company officers must make sure that they are calling for the correct-sized line to knock down the fire. Always pulling the 1¾" preconnected line at every fire is simply foolish and sets the incident up for failure.

I read an account of a big-city fire department that arrived at a well-involved fire in a commercial occupancy. Heavy smoke was showing from the front door as the first-arriving engine company pulled up to the scene. The officer instructed his hose team to pull a 1¾" preconnect, which they took through the front door. Unfortunately, the fire was located at the back of the store and was larger than they expected . . . obviously. The hose team was soon pushed back by the advanced fire conditions, unable to get water on the seat of the fire. The fire soon took control of the situation and burned down the building.

Pulling a small line at a commercial fire is, sadly, not uncommon. Nine firefighters lost their lives a Sofa Super Store in Charleston, South Carolina, in 2007. The NIOSH firefighter fatality report (F2007-18) stated:

> *In this incident, the loading dock area contained approximately 2,300 square feet of floor space, the right showroom addition contained approximately 7,000 square feet, and the main showroom contained approximately 17,000 square feet of floor space. Applying the National Fire Academy rule (area divided by 3), a minimum of 800 gallons per minute (gpm) of water would have been required at the loading dock. Crews operating at both the loading dock and the right showroom addition initially employed 1½" preconnected hand lines capable of flowing 90 gpm. 1-inch booster lines were also deployed. As the fire progressed, 2½" hand lines capable of flowing 350 gpm were put into operation, but their use was hindered by inadequate water supply so that the actual flow rates likely never approached these capacities during the incipient fire stage due to the small diameter of the supply lines.*

Firefighters being “undergunned” with small hoselines—both 1½" lines and 1" lines—was one of many mistakes that contributed to this tragic event.

Fires in large, non-compartmentalized structures usually require large volumes of water to be knocked down and controlled. If you arrive at a commercial building with smoke showing, pull the 2½" line. The 2½" line gives you about 300 gallons of water per minute, and it has a reach much greater reach than the 1¾" line.

The 2½" line is the line of choice for any large fire, or any fire inside a commercial structure, which are oftentimes not compartmentalized like residential buildings (i.e. single-family dwellings and apartment houses). As the late Chief Alan Brunacini told us, “Big fires require big water.”

This brings me to a short discussion of high rise office buildings. High rise office buildings are mostly not compartmentalized, with each floor being thousands of square feet in size. Picture a stack of supermarkets in the sky. This is truly how you should start thinking of a high rise office building. Now picture a fire on one of these floors, say floor 25. What handline do you tell your engine company to pull? Remember, the floor space in a typical high rise is 20,000 square feet or more. Would a fire of any significance in a space like this be controlled with a 1¾" line? Probably not . . . unless you got to the fire at its ignition stage, in which case you would be very lucky. The smart choice is to take the 2½" line and call for more, because with a fire of any significance, multiple 2½" lines will be needed. And even with the 2½" lines, the fire might not be controlled until it has used up the available fuel.

Again, don’t bring a knife to a gunfight! For fires in larger, non-compartmentalized structures, use the bigger line in your arsenal—the 2½" handline. I believe that any engine company worth its salt can become proficient at using the 2½" handline. Have your company drill with this line so that it becomes a tool they are comfortable with and know how to use. That way, when you do have a big fire inside a big structure, your crew will have the confidence and the skills to operate the 2½" line effectively. By regular drilling with the 2½" line, your crew will know how to move it down a smoke-filled hallway. They will know how to handle the backpressure when the nozzle is open. They will know what it takes to move a charged line around a corner. Constant drilling is the key to success. As legendary UCLA basketball Coach John Wooden was fond of saying to his players, “Failure to prepare is preparing for failure.”

All our engines in the Seattle Fire Department are equipped with a 200' preconnected, 2½" handline, tipped with a 1⅛" smooth bore nozzle. Our engines also carry 500' of 2½" hose in a bulk bed, with the first 100' tipped with a 1⅛" smooth bore nozzle. Along with these two lines, the engines carry 200' of 2½" line tied into two separate 100' bundles, which we call a “skid load.” The skid load provides crews with two sections of 2½" hose readily available to charge a standpipe or sprinkler Siamese, to use as a second hydrant supply, or to extend a 2½" hoseline. Our crews are very good at deploying and operating these larger lines because we drill quite often with them as a department.

Stretch Enough Hose to Reach the Fire and Manage the Hose Efficiently

How many times have I seen the first-arriving engine company lay short to the fire? Unfortunately, more times than I should!

The first-arriving engine company must take the time to assess the correct length of hose that needs to be stretched to the fire. However, this doesn't always happen, and the hose team ends up short, yelling for more line over the radio. I've seen an engine company stretch a 200', 1¾" preconnect into a 300'×300' commercial occupancy fire without having a clue as to where the fire was inside the structure.

If you are assigned to the hose team on an engine company, you should make it a habit of doing a quick reconnaissance to see where the fire is and how much hose you might need. If the officer on your engine company is not estimating how long the lay is for your team, then the hose team must—*MUST*—take the time to estimate the correct amount of hose to reach the fire and all the areas impacted by the fire in the structure. If you are unsure about how much line to take with you, then grab an extra length or two of hose. Although too much hose can become a problem as well (where do you put this excess hose . . . inside the structure? . . . outside the structure?) it is much better to have too much hose than not enough. One of the drills I enjoy doing when I am working on an engine company is to drive to a building, point to a location in that building where an imaginary fire is located, and ask the crew how much hose they think it will take to reach a fire at that location. It is amazing the range of estimates I sometimes get during this drill. After everyone has estimated the hose stretch, we then stretch a dry hoseline to the location inside the building. The crew is oftentimes shocked, particularly if they are young and inexperienced, at how much hose was necessary to make the stretch. However, if you continue to drill your engine company this way over time, you will soon see how much better they are at estimating the correct amount of hose to stretch.

I recall a fire one night when I was working as a firefighter on a Ladder Company 5. The call came in as a fire on the first floor of a three-story, wood frame apartment building. Our ladder company arrived just after the first engine company. Flames were blowing out of all the windows of this apartment, including a large sliding glass door to the deck, and exposing the deck of the apartment above.

While my partner, JP, and I masked up at the side door of the apartment house, the hose team pulled a 200' preconnected 1¾" line from the back slot of the engine.

When JP and I forced the door with the irons, we were immediately met by smoke that was banked halfway down the hallway walls. The hose team moved past us with the charged preconnect and advanced down the hallway toward the fire, ducking below the smoke. It appeared that they would have water on the fire quickly.

I was about ten feet behind the hose team when I heard them yelling over their voice amplifiers. JP and I hurried up to their location.

"We're short!" one of them yelled. "We don't have enough hose!"

Before I could say anything, both members of the hose team were running down the hallway back to the engine to get more hose.

I looked at JP through my mask. The smoke was now three-quarters of the way down the hallway wall, and we were on our knees.

"Now what?" I asked.

JP looked down at the charged hoseline and noticed that it was lying on the hallway floor in an S pattern.

"Hey Stevie," JP said. "I think we can go back and stretch this hose a bit tighter and make it just inside the front door to the fire apartment. Let's get rid of these S's. Man, I wish those engine guys had brought another length of hose . . . it's going to be close."

"You're right J," I said. "I'll go back to the side door and straighten it out as much as I can from there."

We worked fast, getting as much slack out of the hoseline as we could. We ended up back at the nozzle with about fifteen feet of extra hoseline.

JP put a strap around the doorknob to control the door, and then we went to work on it with the irons. We popped the door open, but kept it closed with the strap. When I opened the door, I saw that the entire apartment was fully involved with fire. JP grabbed the nozzle and bled the line outside the door, and we advanced inside the apartment, putting out fire as we pushed our way down the hallway. We knocked the main body of the fire down in less than a minute from our position in the front hallway. There was still no sign of the hose team from the first-arriving engine company.

"Man," JP said, "those engine guys are going to be pissed off . . . Truckies knocking down their fire . . . man-o-man . . ."

We both started laughing. How often does a ladder company get to say that they knocked down a well-involved apartment fire? We must have laughed for a minute straight.

I could hardly contain myself when I announced over the radio, "Command from Ladder 5, team B, we have water on the fire. The fire is under control."

"OK Ladder 5, team B," Command responded. "You have water on the fire, and the fire is under control. It looks good from out here."

The moral to this story is this: Make sure you stretch enough hose to reach the fire, and manage that hose by making an efficient and orderly lay. *Take the time to do a quick recon.* If there is ever any question in your mind of how long the lay is, bring an extra length of hose, just in case. It is better to bring too much hose than not enough.

One last point on laying short. If you do happen to lay short and cannot reach the fire location with your hoseline, tell the IC over the radio. If you don't let anyone know that you do not have enough hose to reach the fire, then the back-up line will probably lay short too, since they are probably going to bring the same amount of hose that you did, unless you tell them otherwise. I have seen this happen on the fireground—the first line was short, but since the first-arriving engine company didn't tell anyone they were short, the engine company laying the backup line laid short as well. And meanwhile, the structure fire was burning out of control and extending to the cockloft of a three-story, occupied apartment house. Because the first-arriving engine company didn't tell anyone they had laid short to the fire, the fire grew, extended into the structure, took much longer to extinguish, and caused more property damage.

Again, if you lay short, do the right thing, even if it means that everyone will know your mistake. Tell the IC that you laid short over the radio so the next line doesn't lay short as well.

Water: Don't Ask for It Until You Are Ready

Besides laying short to the fire, another common mistake the first-arriving engine company can commit is charging the hoseline too soon, either by the hose team calling for water too early or by the driver (chauffeur), charging the hoseline before the hose team calls for it.

I remember a room fire on the third floor of a large apartment house. The driver of the first-arriving engine company had charged the hoseline before the hose team even reached the front door—why he did this, I still don't know. I arrived on the second engine company and watched as the hose team from the first engine struggled to move the charged, tangled hoseline into the apartment house and up the stairs. The hose team was having so much difficulty that when my partner and I arrived with our uncharged line and offered to help them, they were more than happy to abandon their line. They helped us deploy our dry line up the stairwell. After we called for water, they then assisted us in moving our charged line down a long, smoke-filled hallway and into the apartment to put out the fire.

Another issue concerning water and hoselines has to do with where an engine company places a charged line in relationship to other lines.

We had a fire in the north end of Seattle where an engine company dropped the back-up line on the first-arriving hose team's line in a stairwell of a six-story apartment fire. When they charged the back-up line, the initial hoseline was lying underneath a mess of charged hose, making it almost impossible for the initial line to be moved anywhere. The initial line had to be abandoned because it was buried under the charged back-up line, which delayed getting water to the fire on floor five.

A better alternative would have been to open the stairwell door to the floor below the fire and flake out the dry back-up line in the hallway. Again, once the first line is in operation, the back-up line can be charged and then moved up the stairwell to "back up" the first line.

Deploying hoseline, particularly in apartments and high rise buildings, is becoming a lost art. I highly recommend that you read *every* article in *Fire Engineering* magazine written by Captain Bill Gustin of the Miami-Dade Fire Department. Gustin is a veteran big city engine officer who explains in detail every aspect of hose handling, deployment, and operation. He truly is a craftsman at his trade, and his writings are always informative and easy to read.

Ladder Company Tactics

Let's talk about the tactics of the first-arriving ladder company. Remember, all ladder company operations "support" the engine company—from locating the fire, to forcible entry, to ventilation, to laddering and overhaul. A good ladder company supports the engine company's main objective of getting water on the fire quickly and makes the engine company's job easier. Everything a ladder company does is done to support the engine—everything!

As I stated earlier in the chapter, the front of the building belongs to the truck—the ladder company. Try to think of the ladder company as a big tool box, and this tool box must be located in front of the building so that the "tools" are close by and can be used. This means that the second-arriving engine company must not enter the fire block until the first ladder company arrives and takes position near the front of the building (if the roads are small and only allow one apparatus at a time). Of course, if the roads are big and wide, as in many cities around the nation, then the second-arriving engine company need not wait—though they still cannot park in front of the building.

Using our sports analogy, what ladder company tactics, or operational plays must be accomplished at every structure fire so that the fire can be extinguished safely and effectively?

Locate the Fire

Before anything can happen on the ladder company, the good ladder officer must be able to locate the fire in the structure quickly, or at least make a good guess where it is, and be able to predict where it might go if it is not brought under control quickly by the engine company. By mentally locating the fire, you will then know who is in the most danger and who needs to be rescued first.

Is the fire in the front or the rear of the structure? What floor is the fire on? Is it in a shaft? Has it extended into the cockloft or attic space? Is there an exposure above the fire? Is there an exposure on either side of it?

Good ladder company officers must be able to look at the cues and patterns that the structure fire gives them and then make the best guess where the fire is located. Oftentimes fire is blowing out a window or door, and the fire location is easy to find. But sometimes it isn't so easy, and all you have is smoke coming from almost every location of the structure. Again, read the smoke, as Dodson says. What is it telling you? By looking at its volume, velocity, density, and color, you should be able to predict, with some degree of certainty, where the seat of the fire is located.

Unless there are known victims in known locations who need to be rescued immediately, or if the structure is unsafe to enter, then first-arriving ladder company must get a team inside to verify where the fire is located and possibly isolate it—more on this latter.

Locating the fire is the most important task of a ladder company when lives are not in immediate danger from the fire. Why? Because if we don't know where the fire is, then the engine company could well be stretching the line to the wrong location, which never turns out well.

Another point on locating the fire—the ladder company team that goes inside the structure to locate the fire can give the engine company a "heads-up" as to how long a lay they will have and what size handline they might need.

I remember listening to a fire one night over the radio while I was working in the south end of the city. Captain Andrus was working downtown on Ladder 10 that night, and a fire came in at a high-rise apartment house right down the street from station 25. The building had long hallways and was made of concrete. The initial report from the dispatcher said fire was blowing out from an upper story window.

When Captain Andrus and his partner, Eddie, made it up to the fire floor, he got on the radio and said, "Command from Ladder 10, team A, we are on floor 11, and have heavy smoke in the hallway with a moderate heat condition. We are making our way to down the hall to the fire apartment."

Several minutes later he came back over the radio.

"Engine 25 from Ladder 10, team A. We are at the fire apartment on floor 11. It is over 200 feet from the stairwell door. You better bring another length of hose with you. We are closing the apartment door to the hallway. No victims near the entry."

Engine 25 did bring another length of hose with them, and therefore had enough hose to make it to the apartment and quickly darken down the fire.

Without Captain Andrus's heads-up call for an extra length of hose, Engine 25 might have laid short, allowing the fire to continue burning unchecked.

Locating the fire supports the engine company, which is what ladder companies do, and it should be at the top of ladder company tactics.

Position the Aerial Apparatus

We have already asserted that the front of the fire building belongs to the ladder company. Now it is time to talk about apparatus position in the front of the fire building.

At every structure fire, the ladder company officer must make sure that the aerial can be used to its fullest potential. If the aerial apparatus is not positioned to make use of its aerial ladder, then why even bring an aerial to the fire? Remember, the citizens of your city or district provided the money so that your department could purchase the aerial apparatus—which is a **very** expensive piece of machinery—with the expectation that the fire department would use it to help protect them.

I have seen countless pictures from around the country of aerial apparatus being placed so far from the fire building that the aerial is simply useless. There are certainly times when the aerial ladder should not be used—if there are electrical wires blocking the front of the fire building, or if it would be faster and easier to ladder the fire building with ground ladders, such as at a fire in a ranch-styled (one story), single-family dwelling. However, if your fire department or fire district is fortunate enough to have an aerial device, it should end up at the front of the structure on fire, and the aerial ladder should be used whenever possible (fig. 4–3).

Figure 4–3. Two firefighters on an aerial ladder.

Force Entry

If the first-arriving engine company is having trouble forcing entry into the fire building, or if they are busy stretching the initial hoseline—which is often the case—then the ladder company must force entry into the structure.

This may involve using the "irons" to force locked doors, using bolt cutters to cut locks or chains securing gates in fences surrounding a building, using the rescue saw to open metal roll up doors, or countless other methods of getting into a secured or hard-to-access structure.

Hopefully we have all been taught to "try before we pry," which simply means that before you begin to force a door or gate—or whatever is blocking entry into the building—you should try opening it first to see if it is already unlocked. This common sense approach is often overlooked by new members to the fire service who are simply too focused on the task to step back and think before they start working on a door or gate.

A word of caution here. If you are still using your foot or shoulder to open doors—Stop! Using any part of your body as a forcible entry tool is a sure way to find yourself on disability. I've witnessed it firsthand several times. I remember seeing a firefighter attempt to kick a locked door in. Unfortunately, his knee suffered the brunt of the damage as he progressively kicked at the door with more and more force. He ended up limping away from the door, which never did budge. That firefighter ended up on disability for over six months with a torn ligament and ripped cartilage in his right knee. The solid wood door he had been kicking had multiple locks, and was hard to open even with the irons.

Forcible entry is a truck skill that requires constant practice and continuing education to know about the latest locking devices and how to overcome them. Recognizing a lock and knowing how it operates are critical components of being able to open them. Truckies must also have a thorough and working knowledge of all the forcible entry tools available to them on their apparatus. This involves constant drilling on these tools, skills, and locks.

I would argue that forcible entry does not end with the opening of the door or gate. Instead, the truly proficient ladder company member should also make sure that the door they just opened stays open. This means chocking the door open with any number of devices—from a simple wood wedge carried in your bunkers or on your helmet, to a fancy store-bought device that serves the same purpose. Of course, if there is active fire behind the door, the forcible entry team should control it until an engine company arrives with a hoseline.

Why chock the door or gate? Truckies should get in the habit of chocking doors so that the door they just forced does not close on the hoseline laid over the threshold. A good ladder company will have chocked the doors all the way to the fire location for the engine company. Again, this supports the engine company as they get water on the fire.

Confine the Fire

A competent interior ladder crew must get to the seat of the fire as quickly as they can, and confine and isolate it, temporarily at least.

How do you accomplish this?

If at all possible, close the door to the fire room. This very simple act isolates the fire and buys you, and the engine company advancing the hoseline, some time. It also helps stop the flow of air—oxygen—to the fire, which will help keep the fire from getting hotter faster. Closing the door disrupts the flow path (see Chapter 1).

Another way to confine and isolate the fire is to use the pump can—the portable fire extinguisher, the Can, or whatever your department decides to call it—that the first ladder company team brings with them into the fire building. This interior team should also bring the Irons, the TIC, and a 6-foot hook with them to the fire location. Use the pump can to spray the ceiling of the hallway to slow an advancing fire overhead, or to knock down the fire enough to buy you some time. In a room fire situation, use the pump can on the fire inside the room and then close the door to the hallway. This confines the steam generated from the pump can water and allows it to put out more fire. Used correctly, a pump can will extinguish a good amount of fire. To use the pump can properly, place a gloved thumb or finger over the pump can nozzle. This serves to break up the water stream from the pump can, which helps distribute the water droplets over a wider area of the material on fire. In some cases, crews have used the pump can to extinguish the fire in the room of origin.

Remember, the first truck team inside the fire building must communicate the exact location of the fire to the advancing hose team if they have not already arrived. Once the proper-sized hoseline is in place and successfully operating, the fire is being confined and extinguished at the same time. This helps all interior firefighting teams and potential victims.

Rescues and Targeted Search for Known Victims in Known Locations

If the ladder company arrives and there are either visible victims in immediate danger from the fire or a credible report of a known victim in a known location, then the members of this Ladder Company must make the rescue or conduct a targeted search of the room or area where the victim is supposed to be.

Picture a person hanging halfway out a window on the third floor of a five-story apartment house. The person is yelling for help and looks like they are ready to jump. Heavy black smoke is pushing out of the top half of this same window under pressure. The ladder company must park the rig and quickly throw a ground ladder to rescue this person. This is a true rescue, one that needs to happen fast or else the person will jump or be overcome by the smoke.

Now picture a woman outside a two-story house. She is on the front lawn as the ladder company arrives. Fire is venting from two windows on the first floor. The woman is pointing to a second story window on the Delta side of the house. Smoke is coming from this open window, and the woman is frantically yelling that her child is up in that room. Members of this ladder company must rapidly throw a ground ladder to that window, get inside the room, and remove the child via the ladder. This is another true rescue of a known victim in a known location.

Ladder company members must continually practice for these type of scenarios, using ground ladders, the aerial ladder, the TIC, and all the various tools they may employ to help them make these types of rescues. Rescues and targeted searches for known victims in known locations are also done by engine company members, but if the ladder company arrives with the engine, then these duties fall to the ladder company members. Why? Because this frees up the engine company to get water on the fire, which makes everything better.

Primary Search

If there are no visible immediate rescues and no reports of known victims in known locations, a thorough primary search must be a top priority of any ladder company working at an occupied building fire. The primary search—or first search—of the fire compartment (room, rooms, or floor) must be focused, rapid, and methodical. And this primary search has traditionally been assigned to the first-arriving ladder company. However, if your fire department or fire district does not have a dedicated ladder company, then this primary search is going to have to be performed by an engine company.

Once the first-in officer has done a size-up that includes a risk assessment (see Chapter 3), and determines that an interior search is warranted, a crew must get inside the structure and get to where the victims are *most likely* to be in peril. This usually means that the primary search should first be conducted on the fire floor and the floors above it. People who are located on these floors are the ones in real danger. People located in horizontal exposures or below the fire are not the people in real danger. Again, the primary search must be focused on those victims or potential victims who are in real danger.

Take, for example, a late night fire that starts in the kitchen, located on floor 1, of a two-story, wood-frame house. Heavy smoke is visible on floors 1 and 2. Where are the victims most likely to be? In the bedrooms, which are up on floor 2. A primary search should begin on floor 2, concentrating on the bedrooms. Remember, it is the smoke—the products of combustion—that is the killer in the majority of structure fires. We also know that most people who perish in house fires are found in their bedrooms, so it would make sense that we concentrate our search efforts there.

How do you conduct a primary search? The interior ladder company crew must quickly get to the fire floor or the floor above it—wherever the victims most exposed to smoke and heat are likely to be. Remember, people located in horizontal exposures and the floors below the fire are secondary to those on the fire floor and the floors and stairwells above the fire. A note of caution here! Ladder company members conducting primary searches above the fire must check in with the hose team so they know the ladder company is on the floor above without a hoseline.

Once on the fire floor or floor above, use the TIC to start a focused, quick, search. The TIC allows firefighters to quickly and effectively search a large area for potential victims. I believe that the TIC is the most important technological advancement to life safety since the advent of the SCBA. If you have a TIC on your rig, you should be using it, and more importantly, training with it—learning its strengths, weaknesses, and limitations. If you are not, then you are doing yourself and the citizens you have sworn to protect a disservice. For those who wish to learn more about the TIC, go to the National Institute of Standards and Technology (NIST) website, which has an informative section all about the TIC.

VENTILATION

Ladder companies are responsible for ventilating heat, smoke, and toxic fire gases (the products of combustion) from the fire building, both vertically and horizontally. Ventilation supports the engine companies, who are extinguishing the fire, and the teams tasked with searching for victims. Ventilation is a *support* operation. Ventilation operations *support* the engine company trying to make it to the fire compartment or floor to extinguish the fire. Ventilation operations *support* the interior ladder company team—or whoever is tasked with the primary search—trying to search the fire floor and the floors above for possible victims.

Ventilation of a structure fire can be accomplished vertically, horizontally, or both at the same time. The type of ventilation necessary depends on several factors: building construction, building type and layout, the location of the fire in the building, and whether the fire has self-vented or not.

Ventilation makes the building behave. Tom Brennan explained ventilation this way many years ago, and it still makes perfect sense today. Ladder companies must make the fire building behave so that the engine company can get inside the building safely and extinguish the fire. In modern terms, Brennan was saying that the flow paths in and out of the structure need to be controlled in a coordinated way. The paths of fresh air entering the fire building, and the products of combustion (heat, smoke, and fire gases) leaving the building, need to be recognized and managed through coordinated ventilation of the building. This means that the days of ladder companies taking out every window at a structure fire are over. Officers and firefighters tasked with ventilation must account for the flow path of the fire and ensure that their ventilation tactics do not increase the risks to the firefighters inside. How is this achieved? By understanding the concept of flow paths and controlling the flow path—by shutting doors, venting the correct window of the fire room when the hose team calls for it, cutting a hole in the roof, and/or using a PPV fan (more on this later).

I remember a fire one afternoon with Engine 6. The report came in as a house fire, with fire visible from the windows on floor 2. Ladder 10 was the second ladder company listed on the response. When we arrived, I saw heavy flames venting out of the floor 2 windows on the Charlie-Delta corner of the house, and thick black smoke issuing from several other floor 2 windows on the Delta side. Engine 6 already had a 1¾" preconnect deployed through the front door. Ladder 3, the first-due ladder, was having trouble deploying their extension ladder due to a severe slope on the Bravo side of the house—their ladder was coming up short.

Just then I heard Mitchell, the lieutenant on Engine 6, tell command over the radio that his team was being pushed to the floor due to the high heat condition on floor 2. He said that they were hitting the fire above their heads, but that they needed ventilation, and needed it quick.

That day I was riding on Ladder 10 with my veteran crew of Jrob, Collin, and Schon, who had all been working with on C-Shift with me for many years. I told them to grab a 35' ladder along with the rooftop ventilation package and place it on the Delta side of the building.

Jrob, Schon, and Collin soon had the 35' ladder up. After a quick buddy check, we all masked-up, began breathing air from our SCBAs, and climbed the ladder. Jay and Schon quickly and efficiently cut the roof open with the chainsaws. They made a big hole near the peak, since the fire was well advanced on floor 2. When they were finished, Collin and I louvered the cut roof sheeting. Then we punched through the ceiling sheetrock with our hooks. Immediately, flames shot out of the vent hole, at least ten feet over our heads. Jay and Schon moved down the roof, cutting the hole larger as they went.

Not long after we finished, we saw white steam and a water stream coming out of our vent hole. Engine 6 was now able to move down the floor 2 hallway and knock down the fire.

Later, when the fire was tapped and we were putting our equipment back on the rig, Lt. Mitchell came up to me and told me how quickly the heat lifted once we opened the hole in the roof.

“We heard the chainsaws, then heard you punching through the sheetrock of the ceiling, and then the heat lifted. One minute we couldn’t move due to the heat . . . and then everything got better. The heat was gone, and we could see,” Lt. Mitchell said. “Tell your guys thanks from me and my guys. Without Ladder 10, we couldn’t have knocked that fire down.”

I told my guys how grateful Engine 6 was for the ventilation hole they cut. Truthfully, a ladder company can receive no greater compliment from an engine company than that they helped the engine company get to the fire. Or, as the late, great Tom Brennan from the FDNY would say, "The ladder company did its job—they made the building behave!"

Books have been written on ventilation tools, techniques, and tactics by experts in the field. I encourage readers to read the books of John Mittendorf and John Norman, and the articles of Mike Dugan and Tom Brennan. For a great discussion of positive pressure ventilation, which I have used throughout my career and firmly believe in, I would direct readers to Garcia, Kauffmann, and Schelble's excellent book, *Positive Pressure Attack for Ventilation and Firefighting.* I also would point readers to ongoing research into both Positive Pressure Ventilation (PPV) and vertical ventilation—see the NIST videos at the NIST website.

A quick word on Positive Pressure Ventilation here. For any firefighter who discounts PPV outright, saying that it is dangerous and does not work, I would ask them if they have ever used it at a fire. And more importantly, I would ask them if they have ever trained with PPV techniques at live-fire evolutions. In my experience, fire departments that truly commit to training their ladder companies—or engine companies—in PPV use it successfully at structure fires, day-in and day-out. PPV works. It is fast and efficient.

Picture a fully involved apartment on fire. This apartment is located on floor 10 of a twenty-story apartment house in a downtown district. Fire is blowing out one window, and heavy smoke is issuing from five or six other windows in the apartment. How are you going to ventilate this apartment fire? Pressurizing the enclosed stairwell and thus the fire hallway with a PPV fan is the quickest, easiest, safest way to ventilate this fire.

Laddering

Placing ladders at the fire building is the responsibility of the ladder company—if your department or fire district has dedicated ladder companies, that is. If your department does not have a dedicated ladder company, the laddering job at the structure fire still must get done.

Ladders provide above-ground access to the fire building and exposures. Ladders give firefighters, and civilians who need to be rescued or removed, alternative entry and exit points to structures. Properly positioned aerial and ground ladders of the correct size have saved both civilian and firefighter lives. These ladders must be placed at the front and rear of buildings, and oftentimes on the sides, and they must be placed correctly—the tip of the ladders beneath the window sill for rescue.

Ladders are sometimes used inside structures on fire. When an interior stair or set of stairs has been weakened or collapses due to exposure to fire in an apartment building, ladder company personnel must place roof ladders or baby ladders over the missing stairs so that the engine company can get to the fire above with their hoselines.

Ladders can also be used to ventilate upper story windows that are beyond the effective reach of pike poles and hooks. Ventilating windows with ladders takes practice and know-how. When it is done correctly, and when coordinated with the interior hose team, it truly is a thing of beauty.

Laddering buildings can be difficult due to terrain, set-backs, slopes, power lines, parked vehicles, parapets . . . you name it. That is why every ladder company officer *must* have their firefighters practice throwing ladders every week. Take your company out on a weekend and throw ladders at a building that is closed for business. Try to find some hard-to-reach areas of buildings and see what size of ladders would work and not work. Take your crew to a drill tower and try to move an extended ladder to an adjacent window on the same floor by rolling it or using the tip-then-butt technique. Get out and drill, over and over, until your crew can throw and move ladders quickly and efficiently, without thinking about it.

Checking for Fire Extension

Has the fire moved into some other area of the structure besides where the main body of fire is located?

Usually, the interior ladder company team looks for fire extension after it has conducted the primary search. With the advent of the TIC, looking for extension has become much easier and faster. The TIC allows firefighters to "see" through the smoke that is typically present throughout the fire room or fire floor (or floors).

But where does fire extend to in a structure? This is a great question, one that has been posed to me by many young firefighters that I have had the honor to train.

The answer to this question is almost anywhere in the fire structure, based upon the type of construction and the location of the original fire.

As I stated earlier in Chapter 2, all structural firefighters must have a working understanding of all types of building construction—from conventional (or legacy) construction methods and materials to current lightweight construction methods and materials, and everything in between. Remember, building construction is the second core principle of structural firefighting.

Why is building construction so important? The simple answer is this—because we work in and around structures of all types. When it comes to fires, buildings are what we do. Buildings are our work environment. Add fire inside a building and now you have a hostile work environment that is trying to injure or kill us. Remember what Frank Brannigan said: "The building is your enemy. Know your enemy."

To understand fire extension, firefighters must also be thoroughly familiar with basic fire behavior. Remember that the first core principle of structural firefighting is fire behavior (see Chapter 1). How fire behaves and spreads is a science, one that structural firefighters must study and comprehend.

So again, where should we look for fire extension? Considering both the core principles of fire behavior and building construction, firefighters must look for fire extension on all sides of the original fire—in any adjacent rooms or areas, in any shafts or pipe chases, in the void spaces of the walls, and above in the attic or cockloft.

Picture a fire in a two-story, wood frame, single-family dwelling. The fire is in a room on the second floor, with flames visible from two windows. The fire has self-vented out the windows and flames are lapping up underneath the soffits. Smoke under pressure can be seen coming from the attic vents near the peak of the roof. Several of the upper story windows adjacent to the fire room also have smoke pushing out of them.

We can be certain that the fire has extended into the attic space. Why? Smoke under pressure—the key here—is issuing from the roof vents. A good interior ladder company would let the hose team knock down the fire in the room of origin, and then use their hooks or pike poles to pull the ceiling down so that the fire in the attic could be extinguished.

We can also deduce that the door of the fire room has been left open to the hallway. Why? Because smoke is coming from the windows of the adjacent room, which can only mean that the door to the hallway of this room is also open. Alternatively, the adjacent room and the fire room are connected by an open throughway or opening. A good interior ladder team would check this adjacent room for fire extension.

But what about the basement? What if this two-story house was an older structure built in the early 1900s and the exterior windows lined up? This is where a thorough knowledge of building construction comes into play. This house might have been built using balloon-framed construction techniques that were popular in the early and mid-1900s, where the stud walls of the house have no fire-stopping and run unobstructed all the way to the attic. Wouldn't it be smart to check the basement for fire, just to rule out the possibility that the fire originated here?

In fact, wouldn't it be a good idea to always rule out a basement fire at every structure fire we go to? Too many firefighters die or are seriously injured when they unknowingly work above a fire in the basement.

I recall a fire I went to one early morning when I was a new lieutenant. I was assigned to Engine 18, working in the Ballard area of Seattle, an older part of the city where the Alaskan fishing fleet is based.

The call came in about 0830 in the morning, just as we were finishing up with roll call in the watch office of Station 18. "Engine 18, Ladder 8, Battalion 4, this is a house fire with flames visible . . . there are multiple calls on this one," the dispatcher said after she gave us the address.

We knew we had a working fire, and everyone bunked up and buckled up. My driver, Jack, who had been in the department for more than thirty years, leaned over and said, "I know that address. It's a one-story house on the corner, with a hydrant right there." Engine 18 led out of the house, with Ladder 8 and Battalion 4 behind us.

As we turned the final corner onto 55 street Northwest, I could smell the fire as well as see the large column of smoke at the end of the block. Jack slowed down as we approached the house, and I could see flames coming from the front bay window and the windows on side B, Bravo. Jack pulled past the house to leave room for Ladder 8 in front. We ended up right on the hydrant on the corner.

The house wasn't set back too far from the street, which meant our 200' preconnected 1¾" line was the one to pull.

"Grab the preconnect and I'll meet up with you by the front door," I told Brian and Bobo—Chris's firehouse nickname—who were my guys in back that day. Brian was just off probation and ready to go to as many fires as he possibly could.

Jack had already hooked up the supply line to the hydrant as I took my SCBA from the compartment and did a quick 360 around the house. It looked like the main body of fire was in the kitchen and the living room in the front of the house. I stopped for a second to talk with Jeff, the lieutenant on Ladder 8. The flames exiting the windows had lapped up underneath the soffits and had extended to the attic—smoke was pushing out from the soffits all around the house. Jeff said that two of his guys were going to the roof, while the other team was going inside with us.

I met up with Brian and Bobo, who had already flaked out the hose in the front yard and were finishing up forcing the door with the irons. I gave a size-up over the radio to Battalion 4, who had taken command, and to the other incoming units. Then I masked up. Brian bled the line while Bobo opened the front door.

The front room of the house was fully involved in flames. Brian hit the fire up high with a straight stream from the front porch. Then he began to move the nozzle in a circular motion around the front room. Soon the fire in the front room was knocked down, and the three of us began crawling inside, advancing the line to the back of the house, where the kitchen was located. Brain led, with me backing him up, while Bobo stayed near the front door and fed us line. We were all on our knees. It was hot and dark. I could hear Ladder 8's chainsaw working on the roof overhead.

As we made our way to the kitchen, Jeff came over the radio and said that Ladder 8 had flames out the vent hole they had cut and were getting off the roof. Brian knocked the fire in the kitchen down easily. Soon the inside crew of Ladder 8 appeared behind me with their pike poles.

"Need some help, Lieutenant?" Darren asked.

"Yeah . . . Darren, Marty, pull that ceiling down so we can get to the attic fire," I said.

They went to work quickly, and soon the entire ceiling was gone. Brian moved in and directed a straight stream over the burning underside of the roof decking and joists. Steam began pushing out the vent hole that Ladder 8 had cut. Soon the main body of the attic fire was out.

"Command from Engine 18," I said over the radio. "Fire under control."

I went over to one of the kitchen windows and opened it to help clear out the smoke. Jeff was outside the window, having just come down from the roof. He still had his SCBA facepiece on, and was trying to tell me something. I shrugged my shoulders to let him know I couldn't hear him. Then he motioned with his hand, pointing below where I was standing. I leaned my head out the window and saw flames issuing from a basement window right underneath me.

I immediately got on the radio and told command that we had fire in the basement and we needed two lines down there. Luckily, Engine 35 and Engine 9 were standing by outside the house. They quickly found their way to the basement stairs and put out the fire.

What went wrong? First, I missed the basement fire on my 360. I was too busy focusing on the flames from the first floor, and the extension to the attic. I failed to see the full picture—the fire in the basement. The fire had started in the basement and had extended up to the kitchen by way of the stairwell. It had then extended to the front room of the house and up to the attic space. Fortunately, Jeff, the lieutenant from Ladder 8, noticed the basement fire burning below us after he came down from the roof and let me know. I was just lucky that the basement fire had not compromised the floor joists in the areas me and my crew were operating on floor 1. We might easily have fallen into the basement fire below if they had failed. Just dumb luck . . . that is all the kept me and my crew safe that day.

What did I learn at this fire? First, that I would always rule out an undiscovered fire in the basement at every structure fire I went to from here on out. Either I would do my own check of the basement, or I would make sure someone did it if I was busy doing something else. I also learned that just doing a 360 wasn't enough. The officer needs to look—*LOOK*—at what is going on. Thinking back on this fire, I know that I jogged quickly around the house, not taking the time to see everything the building was telling me. Sure, I was inexperienced as an officer, but I was also guilty of the same thing as many firefighters, which is focusing on the flames instead of stepping back, slowing down, and looking at the entire picture. It was a mistake I would try very hard never to repeat.

Overhaul

Overhaul begins after the fire is knocked down and under control. Overhaul is a systematic look at the involved structure (or structures) to verify that the fire is completely extinguished. A successful overhaul operation is one where there is no possibility of a rekindle.

Overhaul is typically a ladder company function. The overhaul operation must be methodical, systematic, and thorough to prevent a rekindle of the fire. A rekindle means that companies will have to respond back to the structure hours later when smoke or flames are again seen issuing from the fire building. An effective overhaul solves this problem and the embarrassment that goes along with any rekindle.

The advent of the TIC has made overhaul much easier. Ladder crews tasked with overhaul can now point the TIC at all the possible avenues of fire extension and get a pretty good read if the fire is still burning or if embers have not been extinguished. Notice how I said "a pretty good read." Just pointing the TIC around the fire compartment is not enough. This is where a working knowledge of building construction comes into play. Firefighters conducting overhaul operations who understand building construction can "guess" or read where fire might be hidden in a void space, where it is not completely extinguished. Opening up the structure by removing sheetrock to get to the structural members and void spaces is necessary at times. Let me be clear: I am not advocating taking every room down to the studs after every fire. However, a focused, systematic "opening up" to check for fire extension is necessary. And the TIC can help us focus on where we need to open the structure up after a fire. Using the TIC during overhaul helps us concentrate on the areas that we really need to investigate further.

For me, some of the most difficult fires to overhaul are kitchen fires. Why? Fire or embers have a way of "hiding" in and behind the cabinets surrounding the range, or moving through the venting system onto the roof or into the attic. I learned this lesson early from Gary Angel, who was my lieutenant when I was first assigned to Ladder 5, a busy north-end truck. Lt. Angel explained that he had been to "kitchen fire" rekindles after the truck assigned to complete the overhaul had failed to remove all the cabinets and check the vent duct above the stovetop. Lt. Angel had us take the cabinets down and place them outside on the overhaul pile after every kitchen fire. And we never had a rekindle.

Years later, when I became the lieutenant of Ladder 5, and then Ladder 10, I also made it my policy to remove all the cabinets and inspect the ventilation system after a kitchen fire of any significance. Sometimes my crew would gripe and complain about the work—especially if it was after midnight or they had worked hard at the fire. However, they often found undetected embers or active fire behind the cabinets, in the stud walls behind the cabinets, and sometimes in the ductwork. Later, when they heard of rekindles after kitchen fires in some other part of the city, they would speculate that the truck assigned overhaul had not taken down the cabinets. And oftentimes they were right.

SALVAGE

Salvage has to do with protecting the property of the public we serve. This task has historically fallen to the ladder company, since they carry the necessary equipment to protect property from water damage and fire operations—equipment such as tarps, visqueen (rolls of thick plastic), prosser pumps, etc. Approximately 80 percent of fire loss is not related to the actual flames of the fire. Instead, it is due to smoke; water damage from automatic sprinkler systems and fire operations; and damage done by firefighters pulling ceilings or opening up walls, attics, and void spaces looking for fire extension. While ladder companies might not be able to protect property from smoke damage, they certainly can protect property from water damage and damage done by firefighters looking for fire extension into the structure.

Salvage operations are usually not the first or second priority of the first-arriving ladder companies, but as the fire operations progress and begin to move toward the overhaul phase, salvage operations should occur.

A heads-up ladder company must begin to think about salvage, particularly if they are the second or third ladder truck to arrive at a fire that is close to being under control.

So what should the ladder company assigned to salvage do? They need to survey the scene quickly to see what is necessary. This may involve getting to the automatic sprinkler riser and shutting down the water supplying the sprinkler system that had knocked down the fire. Or they may need to go to the apartments below the fire room, force the door if necessary, and see if water is moving down from the original fire apartment. If this is the case, ladder company members can cover the important property with visqueen or tarps to protect it from water damage—think computers, TVs, stereo systems, furniture, etc.

Salvage might also mean moving furniture, valuables, electronics, files, etc. from one area of the fire space to somewhere else where they will not be impacted by water, smoke, or fire operations.

I remember hearing about a fire in a commercial area of Los Angeles from Chief (ret.) John Mittendorf. He explained that the fire was running a large commercial showroom, and that one of the later-arriving ladder companies at the fire took the time to go into the office spaces and salvage the computers before the advancing fire overtook the office spaces. The fact was that all the customer data and invoices were stored on these computers. After the fire was extinguished, the owner of the company showed up. He was unbelievably happy and very grateful to the Los Angeles Fire Department for saving his files, which, at the time, only existed on these computers. The LAFD's professionalism and heads-up salvage operations saved a company, which paid taxes to the city of Los Angeles, and indirectly paid the firefighters' salaries. This example of the LAFD's dedication to service explains why the LAFD was, and continues to be, one of the best fire departments in the world.

Remember, property protection is part of a firefighter's job description. To do this task well takes time, effort, training, and dedication. Salvage operations may not be as glamorous as rescue operations or putting out fires, but in the end, they are very important to the citizens we serve.

Lighting

At every nighttime structure fire I have ever been to, there has never been enough lighting. The fact is, lighting the fire scene, both from outside and inside, is a safety issue. The task of lighting the interior of the fire building after the fire is extinguished usually falls to the ladder companies—again, since they carry string lights, free-standing halogen lights, fluorescent lights, and extension cords. However, the job of lighting the exterior of the fire scene is both an engine and ladder function.

Lighting on the exterior of a structure fire should happen early. After the first-arriving engine company chauffer has secured a supply and charged the first handlines, they should be thinking about using the rig's exterior lighting system to light up the involved structure. Exterior lighting is often an afterthought, but it truly is important for a number of reasons including scene safety of firefighters and civilians, enhanced visibility, and overall situational awareness.

The first-arriving ladder company on the fireground, which should be positioned at the front of the structure if at all possible, is responsible for using its exterior lights to light up the scene. If you don't have a dedicated ladder company, then the second or third arriving engine company should be tasked with exterior lighting—someone has to do it. And hopefully, every rig close to the structure on fire will have their exterior lights lighting up the fireground.

Lighting on the interior of the structure usually occurs later in the fire, when the fire has been extinguished and crews are beginning overhaul and continuing salvage operations. String lights, halogen lights, and fluorescent lights are brought into the structure and powered by the rig's generator or a stand-alone generator. This lighting is critical for the reduction of overhaul injuries and for the fire investigators, who will be trying to determine the cause of the fire.

Again, lighting is often an afterthought on the fireground. However, heads-up ladder companies and engine companies understand its importance to the safety of our members and to the running of a professional fireground.

Automatic Extinguishing Systems and Fire Alarm Systems

The job of shutting down a sprinkler system, replacing sprinkler heads, and resetting the fire alarm system (if the building has one) usually falls to the ladder company.

Excessive water damage can occur to a structure and its contents if the sprinkler system is not shut down as soon as possible. Most residential and industrial sprinkler heads discharge somewhere around twenty to forty gallons per minute.

Once the first engine company inside the structure has confirmed that the fire has been knocked down and is under control, then the next step should be shutting down the sprinkler system. This often involves finding the sprinkler riser in the basement, manually shutting down the flow, and then draining the system of water so that the popped sprinkler heads can be replaced with functioning ones.

Ladder company members must be trained and familiar with all types of automatic sprinkler systems—wet, dry, deluge, pre-action, foam water, water mist, early-suppression-fast-response (ESFR), etc. Members must possess the skills and confidence to replace sprinkler heads—which means practice. And practice means training! After a fire in a building, the popped heads need to be removed and replaced, and the system needs to be restored so that the building and the occupants are protected once the fire department leaves the scene. Members must also know how to reset the fire alarm system, so that occupants can be warned if another fire—and hopefully not a rekindle—was to break out at the building.

Scenes from Structure Fires

The core principles of structural firefighting come into play on the fireground at every building fire. *Photographs by John Odegard unless otherwise noted.*

Scene 1. A fire that started on a boat in a covered marina with no sprinklers. The first units to arrive found multiple boats on fire as well as a fully involved structure fire over the covered marina.

Scene 2. Engine 25 out on the freeway, extinguishing a fully involved truck on fire.

Scene 3. Using a 2½" line during a large defensive fire in commercial building.

Scene 4. The crew from Ladder 1 uses the ladder pipe to help extinguish a large fire in a warehouse.

Scene 5. Crews about to make entry into a fire in a single-family dwelling.

Scene 6. The first-arriving engine company about to put the first line into operation at a single-family dwelling fire.

Scene 7. A ladder company crew finishes up cutting a vent hole at an attic fire in a single-family dwelling fire.

Scene 8. Engine crews begin a defensive attack on an apartment building fire. This building was under construction.

Scene 9. Engine 17 begins defensive operations on a fire in a building under construction.

Scene 10. Engine 31 sets up to attack a car fire next to a structure.

Scene 11. A ladder company leaving the roof of a single-family dwelling after cutting a vertical ventilation hole.

Scene 12. Engine crews attacking a deep-seated commercial fire with large-diameter hoselines.

Scene 13. Ladder 5 opens multiple vent holes at a single-family dwelling fire.

Scene 14. Fire has extended into the attic space. Interior crews are pulling ceilings on floor 2 to extinguish this fire from below.

Scene 15. A, B, C, and D Shifts, Station 38. (Photo by Capt. Baer and FireDogPhotos.)

Scene 16. Ladder 10 and Aid 25 practice making an offensive heat hole, against construction, on an acquired structure.

Scene 17. The crew from Engine 17 makes a stretch down an alley to attack a detached garage fire.

SAFETY

The fifth core principle of structural firefighting is **safety**. You may not realize it, but we have been discussing safety throughout this book—both civilian and firefighter safety. Going back to figure I–1 from the Introduction, you can see that safety ties the six core principles of structural firefighting together.

I would argue that safety unites and must inform any conversation about fire behavior, building construction, strategy, tactics, and training (which we will talk about in the next chapter).

In this chapter, I will focus mainly on firefighter safety, since it is an issue that is near and dear to my heart. And I am going to concentrate on safety inside and around buildings on fire. Too many of our brother and sister firefighters have been injured or killed at structure fires. Falling into a fully involved basement due to a structural collapse, being caught in a flashover, getting lost and running out of air inside a smoke-filled building, getting seriously hurt or dying when a roof collapses . . . these are just a few of the ways firefighters become injured or die inside and around buildings on fire (fig. 5–1).

I am not going to discuss firefighter deaths due to heart attacks and traffic accidents, even though these are the top ways firefighters die. These are oftentimes preventable deaths that could be solved with effective health and fitness programs and safe driving practices, including seatbelts and rig positioning at roadway incidents. I will also not discuss technical operations, such as high-angle, Haz-Mat, or confined space rescue. These days, firefighters are rarely being injured or killed at technical operations events. Twenty-five years ago, it was a different story. Nowadays, firefighters on technical teams sometimes die during training—mostly in dive and high angle—but not at actual technical events. Instead, in this chapter I want to concentrate on firefighter safety in and around structure fires.

Other than active military personnel deployed to a hostile theater of war, firefighters at a structure fire are working on the most unsafe job site in the world. At structure fires, we are operating in an uncontrolled environment all the time. A building fire is not static—it is dangerous, always changing, and unforgiving.

Figure 5–1. No sprinklers were available before a fire began on a boat in a covered marina. The first fire units found multiple boats on fire as well as a full-involved structure fire over the covered marina.

The fire follows the rules of chemistry and physics. Add to this the always-present force of gravity, which is constantly acting upon the structure on fire, trying to pull it down around us. Throw in injured or trapped victims, and you have a hostile, threatening work place that is attempting to injure or kill you as you try to save others and protect their property.

Structural firefighting is a dangerous job. Civilians die in structure fires. And sometimes, even when all known safety precautions are followed and heeded, firefighters die in structure fires. The only way to prevent all firefighter deaths at structure fires is for firefighters not to show up, or, after firefighters arrive on scene, to let every structure burn to the ground. But that is not what we do, and that is not who we are. And that's not what the public who pays our salaries and trusts us with their lives expects us to do. We are there to protect lives and property. And sometimes, even if we perform our jobs as safely as possible, we may pay the ultimate price in attempting to save the life of a civilian.

We all need to remember that fighting fires in structures is not a job for the faint of heart. It is hard, and dangerous, and brutal. But in my opinion, it's the best job in the world.

Fire Behavior

I believe that the most dangerous, and the most common, fire behavior event that firefighters face in structure fires today is flashover. Remember from Chapter 1 that flashover occurs during the growth stage of the fire, when heat is being produced by the fire. Some of this heat is present in the heated smoke and fire gases that rise from the fire to ceiling level and then bank down to floor level. Some of this heat is radiated back from the walls of the compartment and transferred to the contents of the room (this is sometimes called re-radiation). The temperature of the contents rises, causing these contents to begin off-gassing ignitable vapors, which are unburned fuels. The ability of the contents to absorb the transferred energy of the fire is determined by many factors, such as their physical states and their ignition temperatures. And since a fire in a compartment can produce very high heat levels (upwards of 1500°F), it is possible for all the contents of the compartment to quickly reach their ignition temperatures. Because of the quickly rising heat in the compartment, all the contents are now giving off ignitable gases—or *fuel*. Flashover occurs when all the fuel present in the compartment—all the accumulated ignitable smoke, gases, and the exposed combustible surfaces of all the contents—ignites suddenly. Flashover is typically, but not always, full room involvement with fire.

Actual flashovers are sudden and can take place in seconds or in a fraction of a second. In a flashover, the temperature of the fire compartment increases immediately to over 1800°F at ceiling level. These conditions are immediately untenable for firefighters. Firefighters who are caught in flashovers will be either seriously injured or killed.

The good thing about flashover events is that they are usually preceded by some observable indicators or signs. The first sign is commonly described by firefighters as "heavy smoke under pressure" issuing from the structure—a large volume of dense black or dark brown smoke, moving or boiling out of the structure with velocity.

Firefighters met with a pre-flashover scenario should work on cooling the space *before* entering it. This may mean sitting at the doorway or window with a hoseline, directing water at the ceiling to cool it down and keep the flashover event from occurring. The Swedish Fire Service has done extensive studies on flashover and pre-flashover events, thanks in large part to the groundbreaking work of Krister Giselsson and Mats Rosander in the 1970s.

Giselsson and Rosander, both fire protection engineers who taught at the Swedish Fire School in Stockholm in the 1970s, wrote *Fundamentals of Fire* (1978). In this book they looked at the science of fire behavior and recognized the combustibility of smoke in compartments and how this led to flashover events.

Together, they developed new nozzle techniques to cool the super-heated fire gases and smoke in compartment fires, thus preventing flashovers. They advocated brief nozzle pulsations or "bursts" using small droplets of water directed at the ceiling, which would cool the smoke, keeping it from igniting—preventing flashover—without creating steam that would scald the firefighters operating in the compartment.

At the time, Giselsson and Rosander's ideas were a radically new approach to preventing flashover in structure fires. Over time, and with the weight of good science behind them, their ideas began to be accepted and adopted around the world. Paul Grimwood of the London Fire Brigade took the ideas of Giselsson and Rosander and put them to the test in London during the 1980s and 1990s. Grimwood combined the Swedish water fog techniques with tactical venting actions, and came up with his *3D Firefighting* tactics for structure fires.

Thanks to recent NIST studies of flashover in structures, the American Fire Service is slowly beginning to adopt and modify the tactics and techniques of Giselsson, Rosander, and Grimwood. The fact of the matter is that most flashovers that occur in structure fires after fire crews are operating inside the structure can be prevented. And by preventing flashover from happening, we can also prevent needless firefighter deaths and injuries around the United States. If you have never heard of Giselsson, Rosander, and Grimwood, I would highly encourage you to read their body of work.

I remember watching a video of a crew of four firefighters bailing out a second story window of a private residence just as the bedroom they were in flashed over. It was obvious from the smoke issuing from the windows of the bedroom that it was about to flash over—the smoke was thick, black, and boiling out of the windows under heavy pressure. These firefighters were searching above the main body of the fire without a hoseline—perhaps one of the most dangerous situations in which structural firefighters can ever find themselves. They all bailed out the bedroom windows to a porch roof, where a heads-up engine chauffeur had hastily thrown a ladder. Unfortunately, one of the firefighters bailing out of the window fell and broke his leg.

But what could this crew have done differently to avoid this flashover? One thing would have been to close the open bedroom door to the second floor hallway—in other words, *control the flow path* of the fire. This simple act would have delayed the flashover. Having a hoseline with them would also have averted the flashover—they could have applied water to the overhead and kept the products of combustion from flashing. Another thing they could have done was to call for an outside line to direct water into the bedroom through a window. This act would have most likely prevented a flashover by cooling down the overhead. If you drop the temperature in a room by cooling it down, flashover cannot occur.

NIST and UL studies have shown that cooling a room down with a hose stream from outside the structure will temporarily keep the fire from flashing by dropping the interior temperature.[1] These studies also have shown that the fire will not be "pushed" to other compartments in the structure. In every experiment, the situation was improved by applying water from the outside—no flashover, no "pushing" fire into other compartments of the structure.

So what are the take-away lessons and applications for structural firefighters from these important fire behavior studies?

Perhaps the most important take-away is that aggressive firefighters should seriously consider some "new" tactics for battling significant structure fires. Why not take the punch, or energy, out of an advanced fire by quickly cooling it down from the exterior *before* moving inside to extinguish the fire? Some large metro fire departments will commonly apply water from the outside of a structure on fire to "reset" the fire. This keeps the fire from growing unchecked while the interior line is being stretched for an interior attack. Sometimes they do this with a deck gun and sometimes with a hand line, depending on the size of the fire.

I understand that many firefighters out there have been taught that putting any water on a fire from the outside is wrong, that this is not how aggressive companies fight fires in buildings. However, in light of the UL and NIST studies I just mentioned, I think that a thorough review of our current accepted tactics is necessary.

Don't get me wrong. *I'm not saying that firefighters should apply water from outside all the time, at every fire*. What I am saying is that firefighters should consider this approach in those situations where there is a good chance of a flashover, when thick, black smoke is boiling from the structure under pressure—the tell-tale warning sign of imminent flashover.

I remember a fire early in my career when I was working at Station 17, in the University district of Seattle, where the University of Washington is located. I was assigned to Engine 17 that day, a hard-working Engine Company with one of the most experienced captains riding in the front seat that shift, Captain Carlson. Around 1100 hours, Engine 17 was dispatched, along with Ladder 9, with the balance of a full response to a boarding house fire, where heavy smoke and fire was reported showing from the second floor.

When we arrived, we saw flames from two windows on floor 2, auto-exposing the windows above them on floor 3. Heavy smoke under pressure was also moving out of multiple windows on floor 2, telling us that the door to the fire room was open.

Captain Carlson told Monty, our driver, to pull past the front so the truck could have the address. There was a hydrant on our side of the street, right on the corner, which we parked beside.

Captain Carlson looked back at me and Jimmy, my partner, as the rig stopped. "Pull the manifold to the front door, and take 200' of 1¾" off it to the fire. You'll have plenty of hose to make the stretch," he said to me and Jimmy. "And watch out for conditions in the hallway on 2 . . . it looks like it could flash up there."

Jimmy got the manifold, which was supplied by 4" hose, and yarded it to the front door, while I grabbed a bundle of 1¾" hose. I took the hose bundle inside the first floor, just at the landing to the stairwell, and then took hold of a couple of flakes and spread them outside the door so that we would be able to easily advance up the stairwell. Jimmy hooked up the hose coupling to the manifold and made sure the gate to our hose was open. Then we both masked up and did a quick ready-check, making sure that we had our radios on the right channel, we had a set of irons (we did not have TICs on the rigs at this time), and our bottles were full of air. Jimmy went outside and gave a thumbs-up to Monty, who then charged our line with water. The line jumped to life as the water ran through the hose. I bled the line of air and looked back at Jimmy. We were ready.

We quickly made our way up the stairs, opened the hallway door, and were met with heavy smoke, which was banked down halfway down the wall of the hallway. Looking under the smoke, I could see that the fire room was about sixty feet away, with the door open to the hallway. Flames were rolling down the ceiling of the hallway, mixed with the smoke, a classic rollover condition.

"Hit the overhead as we move down the hallway," Jimmy said in my ear. "We don't want it to flash over us."

"Got it." I opened the nozzle for a few seconds, directing a straight stream at the hallway ceiling ahead of us. I noticed that no water came back down, which meant that it was really hot up there, so I opened the nozzle up again and began moving forward, penciling the ceiling in front of us to cool the overhead.

As we crawled toward the fire room and advanced the hoseline, I would stop every five feet or so and open the nozzle for a time, cooling the hallway ceiling in front of us. The flames that we had seen in the smoke disappeared.

When we reached the apartment, I opened the apartment door all the way and saw that the living room and kitchen areas were fully involved in flames. I operated the nozzle from the doorway for a minute and darkened down the fire. Then Jimmy and I moved inside to finish it off.

"Command from Engine 17 team B, we have water on the fire," I heard Jimmy announce over the radio behind me.

We pushed in farther. The sofa was still flaming, as were the cabinets in the kitchen. We moved methodically, putting water on the sofa, and then making our way to the kitchen. As we crawled along the floor, we kept sweeping the floor to see if any victims were present—but we didn't find anyone. We spent some time putting out fire in the cabinets over the stove and then extinguished the refrigerator, which was also on fire.

Ladder 9 members had followed us in the apartment, searching the various rooms as Jimmy and I kept working on the fire. Soon we had the fire out in the apartment.

"Command from Engine 17 Team B, we have the fire under control," Jimmy told the IC over the radio.

I heard Engine 22 on the radio. They had a back-up line off the manifold and were on their way up to floor 2.

"Not much for Engine 22 to do," I heard Jimmy laugh through his mask. "Tough luck for those guys."

I laughed too, feeling the feeling only firefighters know when they knock down a fire by themselves before the second line arrives. It's a feeling of accomplishment, of pride in a job well-done, of knowing that you had done something that many others simply do not get the opportunity to do—and it felt good.

Looking back, Jimmy and I had done exactly what we had been taught to do when confronting a possible flashover situation inside a structure—cooling the overhead with the hoseline before advancing. I didn't know it then, but I was using the short-burst technique described by Giselsson and Rosander to prevent a flashover from occurring.

Another important take-away is that you cannot push fire inside a structure with a straight stream. UL and NIST conducted numerous experiments in structures, some of them held in New York and Chicago. In all instances, the science shows that fire streams do not push fire.[2] Water puts out fire—in every case, once the application of water began, the temperatures decreased inside the structure. And more importantly, no temperature spikes were observed in any of the rooms of the structures involved in the test fires, especially the rooms adjacent to the room where the fire originated.

Applying water from the outside of a structure fire is an important tactic for fire departments and fire districts that show up at a fire with minimal staffing. Now there is scientific evidence that quickly applying water from the outside can free up limited personnel to complete other important tasks, like forcible entry, search, rescue, and ventilation.

Building Construction

Buildings, no matter how they are constructed or what materials are used to construct them, are the enemy of firefighters—as Frank Brannigan told us throughout his long and distinguished career. "Know your enemy" was Brannigan's mantra and his challenge to the fire service. Brannigan wanted firefighters to know how buildings are put together—what the structural elements are made of, what connections are holding the structural elements together, what forces and loads are working on the building, etc.

Firefighters are killed every year by buildings that collapse due to fire weakening the structural elements or the connections that support and transfer the loads of the structure. These collapses can be either partial collapses—like a section of a church roof or a portion of a masonry wall—or these collapses can be total. Either way, firefighters must be able to look at a building that is on fire, understand how it is constructed, and forecast how the fire will weaken or damage the structural elements and the connections that support and transfer the loads of the building.

Firefighters need to get in the habit of "undressing" buildings on fire in their mind to "see" how they are constructed. And this skill takes practice.

Firefighters need to train how to recognize the structural elements and connections that lie beneath the outer appearance, or skin, of the building. They need to learn how to identify legacy (older) construction and newer (lightweight) construction. This skill can be learned, but it takes time, a dedication to understanding building construction, and planning. Officers need to train their firefighters on this valuable skill.

But how? After having their firefighters read about building construction, company officers *must get out of the fire station* and drive around their districts. Out in their districts they can stop at different buildings and point out examples of different types of construction. Company officers can also visit new construction sites and walk through these sites with their crews, pointing out the latest types of materials (such as lightweight sheeting and decking), structural systems (such as engineered flooring and truss systems), processes being used by the construction industry, and the connections that hold the various structural elements together.

Company officers have a duty to learn, master, and then teach everything they know about building construction so that their firefighters can operate safely at structure fires. Company officers should also drill their crews on the firefighter line of duty deaths and close calls where building construction played a part in their death or injury. Great examples can be found at the NOISH and firefighterclosecalls.com websites (both of these website addresses can be located at the end of this book).

Firefighters need to learn the lessons of the Hotel Vendome collapse in Boston in 1972 that took the lives of nine Boston firefighters and injured another eight;[3] they need to be familiar with the roof collapse of a church in Indiana that took the life of a forty-year-old career firefighter in 2011;[4] they need to read about the two firefighters who died and the twenty-nine other firefighters who were injured in March of 2004, when a church bell tower collapsed during overhaul operations in Pittsburgh, PA.[5]

What I am proposing is that firefighters learn about their jobs using the Case Study Method. In his excellent book, *Fireground Strategies*, Anthony Avillo argues that "there is possibly no other profession where case study is more valuable than in the fire service." And I absolutely agree with Avillo. Using the case study method allows firefighters who were not at an incident to experience it, review it, and learn from it. Avillo brings home this point when he talks about how, in Jersey City, they had fire in a cold storage warehouse which was very similar to the fire that occurred in 1999 in Worcester, Massachusetts, where six firefighters tragically perished. But because the Jersey City Fire Department had studied the Worcester, Massachusetts case, they understood the dangers of a deep-seated fire in a maze-like, windowless structure with cork-lined walls. They opted to fight the fire defensively, from the exterior, and suffered no firefighter injuries or deaths.

The fact is that firefighters need to study the many case histories where building construction played a part in a firefighter injury or death. If we fail to teach our firefighters about building construction and how building materials, structural elements, truss systems, and connections have all played a part in firefighters getting injured and killed, then our younger firefighters will not learn the lessons of the past and are bound to repeat mistakes that have already cost firefighters their health and lives.

STRATEGY

The strategy of a firefight is the overall plan. This plan should be based upon accomplishing the three main strategic priorities of Life Safety, Incident Stabilization (fire containment and ultimate extinguishment), and Property Protection. We also talked about how this strategy, or plan, of fighting a fire should answer some basic questions, such as:

- How is this structure fire to be contained, offensively or defensively?
- Where should the firefighters be deployed at the fire?
- How should the apparatus that show up at the fire be utilized at the scene?
- Should the first ladder company to arrive be used for victim search or ventilation?
- Should the equipment and tools be used to extinguish the fire first, or to protect and save any human lives that may be in danger?

We also discussed that the IC—often the first officer to arrive at the fire—must conduct a thorough size-up of the structure fire *before* they decide upon a strategy.

A thorough size-up before deciding on a strategy at any structure on fire is a safety issue. When a good size-up has not been done, things often go wrong at a structure fire, and sometimes, unfortunately, firefighters lose their lives.

Let me say that again, because it is so very important: *Failure to do a thorough size-up before deciding on a strategy at any structure on fire can cost firefighters their lives.*

Unfortunately, my fire department learned this lesson the hard way back on January 5, 1995. That was when four of our brothers died in the Mary Pang warehouse fire, which was intentionally set by the arsonist and murderer, Martin Pang.

The truth of the matter is this: the IC failed to do a thorough size-up of the building. The fire was set by Martin Pang, the son of the owners of the business. He started the fire in the basement of the building, which was not visible from the Alpha and Bravo sides of the building since it was built on a grade. From sides Charlie and Delta of the structure, everyone could see the basement and floor 1 above it, which fronted the streets on sides Alpha and Bravo.

There was confusion on the fireground as crews from Engine 13, Engine 10, and Ladder 7 battled the fire on floor 1, unaware that a fire was raging below them. A failure by the IC to do a complete and thorough size-up—one of the most fundamental fireground responsibilities—played an important role in four of our firefighters perishing in that fire that night.

As the US Fire Administration Technical Report on the incident noted,

> *In retrospect, it can be determined that very different interpretations of the structure and the fire were being made from different vantage points. The individuals making these observations each believed that their interpretations of the structure were accurate, but did not recognize the significance of their information to the Incident Commander.* **This emphasizes the value of a complete 360 degree size-up of the fire scene, as early as possible, by the Incident Commander or by an individual who can report in person to the Incident Commander with a "full picture" of the scene.**

Firefighters need to always keep their focus on the basics of structural firefighting, and sizing up a fire in a building is one of these basics. A thorough size-up sets the stage for the entire fire fight. You don't have to be a Company Officer to do a size-up. Every firefighter on the fireground should be conducting their own, independent size-up, looking at the big picture.

Tactics

Recall that tactics are the actual hands-on operations that must take place in the correct sequence to successfully accomplish the overall strategy of the fire-fight. Tactics are the physical actions engine and ladder companies do, from pulling attack lines to ventilating structures.

These hands-on operations, which are conducted by firefighters on the fire-ground, must be performed with safety in mind at all times. However, for these actions to be accomplished safely, firefighters must be trained on how to conduct these operations safely *before* the fire occurs. This means that firefighters need to be trained on how to safely and effectively operate the equipment they will use during fire suppression activities—different sized hose, nozzles, manifolds, engine pumps, ladders, aerial ladders, tower ladders, chainsaws, rescue saws, forcible entry equipment, vehicle extrication equipment, hooks, positive pressure fans—and how to do so in a coordinated and timely fashion. In the next chapter, I will be discussing the training of firefighters much more thoroughly.

I recall a house fire one night in a north end neighborhood of Seattle. The driver of the first-due engine company that night was not the usual driver on that shift, but the back-up driver. I was on the first-due ladder company, Ladder 10, coming from the Capitol Hill neighborhood.

When we arrived, I saw the engine company driver running around the rig, getting a supply from the hydrant right across the street from the house on fire. He seemed a bit confused and hurried, and I noticed it. The firefighters from this engine were already inside the front door of the house, putting water on the fire.

Several minutes later I was inside the house, beginning a search of the upstairs bedrooms, when I heard the hose team from the first-due engine company come over the radio.

"Urgent . . . Command, we've lost water, we've lost water!" There was panic in the voice of the member speaking. "We are backing out!" At this point, I knew that we were above the fire with no hoseline below us—which is a very dangerous position to be in. My partner and I began to back out as well.

Luckily, Hallzy, who was riding on Aid Car 25 that night, recognized that the back-up driver on the first-due engine had mistakenly put the hydrant supply into a rear discharge port, not the intake. The hose team had used all the tank water on the engine, and since there was no hydrant supply coming into the pump, the engine ran out of water and the pump began to cavitate. Hallzy quickly recognized the problem, shut down the hydrant, reconnected the supply hose to the correct intake port on the engine, and re-opened the hydrant. Soon the interior hose team had water again and, with the help of a charged back-up line, extinguished the fire.

How often do you think the back-up driver on that first-due engine company drilled on the basics of pump operations and hydrant supply? Probably never! And it showed at this fire. I don't blame the back-up driver for this, but I blame the officer who never drilled the back-up driver. Luckily, due to the heads-up action of firefighter Hallzy, no one got hurt and the fire was extinguished.

Safety dictates that fire officers train their crews on all aspects of their job. If this back-up driver had spent time drilling on the basics of taking a hydrant and getting a supply, this type of problem would not have occurred. Officers owe it to their crews, and to all the other crews they work with, to make sure the firefighters under their charge can perform the basic tactics of firefighting in a capable and safe manner.

Another fire comes to mind when discussing safe tactics. Under the tactic of forcible entry is the task of opening a locked door, usually for entering a space where a fire is located, or for an emergency egress for fire companies working inside a structure.

This fire occurred in a parking garage under an apartment house—it was a car fire that had extended to a couch which a tenant had brought down to the parking garage in order to dispose of it later. When the ladder company arrived, the only access into the garage was through an outside man-door, which was locked, or the roll-up door, which was closed. The first-due engine company had decided to go through the man-door and was pulling their hose while the forcible entry team from the ladder company began to attack the man-door with the irons.

The ladder company firefighter working the Halligan decided to use the "baseball bat swing" technique to place the pike end of the Halligan in the jam before forcing the door by prying the bar downward. Unfortunately, the firefighter swinging the Halligan was not paying proper attention, and he drove the pike end of the tool into the hand of his partner, who should never have had his hand anywhere near the jam in the first place. This resulted in a firefighter receiving a severe injury to his hand and delayed getting water on the fire. Fortunately, the firefighter who took the pike end of the Halligan to his hand is back to work today. But this incident shows that both the firefighters from the ladder company were not training, or not training correctly, on how to use the irons safely and effectively.

As a result of this incident, our Operational Training Group decided to build some forcible entry props, and in the course of the next year, we had every one of our firefighters learn—or relearn—how to safely and correctly use the irons and the rescue saw on various forcible entry scenarios.

The above incidents show us that fire officers must continually train on fire tactics and tasks with their crews. This training has to be constant and never-ending. There are just too many tactics that we ask our firefighters to perform, under the most stressful and deadly conditions, for us not to continually train. And this training has to stress the importance of performing these tactics safely, efficiently, and with coordination. If we forget about safety, our firefighters might become part of the problem on the fireground instead of part of the solution. It is as simple as that.

Again, I'm not saying that we should embrace safety to the point of not doing our job on the fireground of protecting life and property. **NO!** Our job is inherently dangerous. But to argue that there is no need for safety on the fireground is disingenuous at best and dangerous and naïve at worst.

Situational Awareness

A topic related to tactics and working inside structures on fire is keeping yourself and your crew aware of what is going on around you. We call this *situational awareness*, and it is not necessarily an easy thing to master. Situational awareness is knowing exactly where you are in the structure relative to how you entered it and where the main body of the fire is located. Situational awareness is knowing what the fire and smoke conditions are inside the structure and if conditions are changing for better or worse during your time in the structure. Situational awareness is knowing how long you have been in the building and how much air your crew members have left in their cylinders. Situational awareness is knowing where a secondary exit is located should conditions deteriorate. Situational awareness is knowing if the crews on-scene are winning or losing the firefight.

Being aware of your situation inside a building on fire is a safety issue. Situational awareness is critical to structural firefighting and can be taught. However, in my experience, it takes many, many hours of realistic, scenario-based training—and working at actual fires—to get firefighters to really figure out what is going on around them during a fire. Most firefighters are task-focused and have a hard time taking in all the information of what is happening around them.

Situational awareness in a structure on fire is a skill that takes time to understand and to master. A good officer can help their crew by taking the time to point out the obvious and the not-so-obvious cues and details at every fire call.

Captain Pat Lucci, one of my first officers when I rode on Engine 31 in the north end of Seattle, took the time to teach me much of what I know today about fire behavior and safety inside structure fires. I remember one fire in particular—we were at a house fire that had started in the kitchen on floor 1 and had moved out into the living room and down the hallway. The owner was out in front of the house when we arrived and told us that no one was inside. He had accidentally left a pan of food on the stovetop and had returned from a neighborhood walk to find his house on fire. Flames were already venting out of two of the living room windows.

After he had walked around the house and made his size-up, Capt. Lucci instructed me and my partner, Joe, to grab the 1¾" preconnect and meet him at the front door. Joe and I forced the front door with the irons while controlling it with a strap around the knob. Soon our driver, Bubba, had charged the line and Capt. Lucci was behind us. He made sure that we had our gloves and hoods on and that we were masked up and ready to go.

"OK," he said, "Joey, I want you to open the door, but I want you both to wait a second and watch the smoke before you go inside."

Joe pushed the door open and heavy black smoke began boiling out at the upper part of the door frame.

"See how fast that smoke is coming out, and see the pressure it's under?" Capt. Lucci asked in his always-calm tone. "Look below the smoke. See how clear it is down there, and how you can see the fire in the living room." I nodded. "We need to cool the overhead down a bit so it doesn't flash on us, and then we can move inside."

Joe and I took note of everything the Cap was saying. We watched the smoke coming out of the front door for a bit longer.

"OK, let's hit it with a straight-stream at the ceiling, about ten feet in, for a few seconds" Capt. Lucci said.

I opened up the nozzle and did as he instructed. In a very short time, the smoke color had changed. It was lighter.

"OK, let's move in low and start with the fire in the living room and work our way to the kitchen. Stevie, keep that nozzle moving in a circular motion, just like we drilled on last week." Joe and I moved in, followed by the Cap. We moved in and started knocking down the fire. And all the while, Captain Lucci was teaching us, showing us what the fire was doing, how it was reacting to the hose stream, and how we could do our job better and more effectively.

Captain Lucci was really teaching Joe and me about situational awareness in a structure fire. He was getting us to take the time to look—really look—at what was going on around us. And he did this at every fire. He was passing on his fire knowledge to us, showing us that we had to be constantly monitoring and paying attention to what was going on around us. He was teaching us situational awareness.

One last word on situational awareness . . . When my partners—Phil Jose, Mike Gagliano, and Casey Phillips—and I started teaching firefighters around the United States how to manage their air in their SCBAs more than a decade ago, we noticed an unintended benefit that arose by forcing firefighters to look at their pressure gauge on their SCBA every so often. When firefighters stopped the task they were doing—like pulling an attack line or searching an apartment—and took a few seconds to look at how much air they had left in their SCBA cylinders, they also began to look around at the environment they were in. Firefighters began noticing more of what was going on around them—the smoke and heat conditions, where they were in the structure, and what was happening with the fire fight. In fact, we discovered that by teaching firefighters the skill of air management, we were also giving them a chance to increase their situational awareness inside the fire building. We never guessed that by managing their air, firefighters would also be paying more attention to their situational awareness—it was an unintended benefit of air management that none of us could have ever predicted.

Complacency

In structural firefighting, complacency can be deadly.

Complacency in structural firefighting occurs when firefighters have become too comfortable doing the same tactic or task over and over again, and they stop paying attention to what is really happening around them. In other words, they get lazy and let their guard down. I have another term for complacency, and that is *operational laziness*. In structural firefighting, operational laziness—complacency—can lead to serious injury or death on the fireground.

I remember reading about a firefighter in big-city fire department who fell off a flat roof of a three-story, ordinary apartment house during a fire and was severely injured. He could not see the roof due to the heavy smoke conditions, and did not drop down and crawl or sound the roof ahead of him. Instead, he walked across the unseen, smoke-obscured roof and right off the side of the building. Luckily, he did not perish, but his firefighting career ended that night.

Or how about the firefighter who wasn't wearing his SCBA during fire overhaul? A fire had been extinguished in a commercial occupancy by the sprinkler system in a large west coast city. When the fire department arrived, firefighters found that the fire was already extinguished by the sprinkler system and that the smoke had mostly cleared from the structure. Several of the firefighters decided not to use their SCBAs to overhaul the fire. No CO readings were taken from a gas monitor before overhaul operations began. This was a poor decision based mainly on complacency—fire out, light smoke condition. What could go wrong?

It turns out that a lot could go wrong. One of the firefighters assigned to replace the fused sprinkler heads began to feel dizzy and soon passed out. He was rescued from the structure by another firefighter and was found to have been exposed to high levels of carbon monoxide (CO) and hydrogen cyanide (HCN). He was lucky, since he was rescued before too much CO bound to the hemoglobin on his red blood cells, and since there was a hyperbaric chamber at a nearby hospital. Other firefighters in the structure who were not using their SCBAs also experienced CO exposure and suffered various symptoms, though not as severe as the firefighter who had first become unconscious.

This incident brings home the point: every firefighter doing fire overhaul **MUST** *wear and use their SCBA* and **MUST** be wearing their full PPE. There are just too many toxins and poisons and carcinogens in the post-fire environment. Failure to do so is falling into the trap of complacency.

Here is another example of operational laziness. A professional, big-city firefighter decided not to use his SCBA while he was venting a commercial roof—and no one up on the roof with him made him go back down the aerial and get his SCBA and face piece. Unfortunately, he later fell through the roof and suffered career-ending burns to his face and airway, as well as some serious fractures, all because he was too lazy to use his department-issued safety equipment.

How many times have you heard a firefighter make the following statements?

- "We never use our SCBAs during overhaul, and nothing bad ever happens."
- "Do we really need to wear our eye protection on this EMS call? It's for a man down on the sidewalk."
- "We don't need to bunk up for this automatic fire alarm at this building. It's just another false alarm, just like the twenty times before this one."
- "I never wear my gloves when I'm using the rescue saw or extrication tools and nothing bad has ever happened to me."
- "Our truck company doesn't wear our SCBAs when we go up to the roof to cut vent holes. They just get in the way."
- "It's just a dumpster fire, so why do we need to put on our SCBAs?"
- "This is just a small fire, so we don't need another ladder to the roof."
- "In the twenty years I've been in the department I've never used my seatbelt—ever—and I'm OK."
- "Do we really need to throw the ground ladders again? We just did that two months ago."
- "We are a veteran crew, so we don't really need to do much hands-on training."
- "Why do I have to put the rig in pump at the start of every shift?"
- "We don't really hold a roll call on this shift since everyone knows what they are supposed to do already."

All these statements and questions arise from complacency. The fact is that these firefighters are suffering from operational laziness. And it is the officer's duty to guard against this complacency trap. Officers must stay vigilant and expect the unexpected, on *every* call.

Are you making sure your crew is buckled up every time they get on the rig? Do you drill constantly on all aspects of your job? Is your engine company ready to problem-solve and overcome a burst section of hose? Can your ladder company cut an offensive heat hole and a defensive trench cut safely and effectively? Can you stretch, operate, and move a 2½" attack line efficiently? Can your crew deploy and raise all the ladders that you carry on your rig with precision and speed?

Fighting against complacency is not easy. In fact, it is often very difficult. But company officers owe it to their crews and the citizens they serve to do the right thing and keep everyone on their rig combat ready—part of which means being constantly on guard against operational laziness.

Beware of the complacency trap—it is unforgiving and can be fatal.

Endnotes

1. https://ulfirefightersafety.org/research-projects/impact-of-fire-attack-on-firefighter-safety-and-occupant-survival.html
2. https://www.nist.gov/el/fire-research-division-73300/firegov-fire-service
3. https://www.firefighternation.com/2012/06/17/hotel-vendome-remembering-the-worst-firefighting-tragedy-in-boston-s-history/#gref
4. https://www.cdc.gov/niosh/fire/reports/face201114.html
5. https://www.cdc.gov/niosh/fire/reports/face200417.html

TRAINING

As I stated back at the beginning of this study of structural firefighting, **Training** is the foundation upon which the other five core principles stand (fig. 6–1). Meaningful and thorough training on fire behavior, building construction, strategy, tactics, and safety must start in probie/recruit school and continue until the day you retire. It is my firm belief that training is the most important core principle of structural firefighting.

Officers must train their crews every shift. They have an obligation to help their firefighters learn and practice their craft so that they can become proficient, knowledgeable, and safe while working at structure fires. There is no "do-over" on the fireground, no "time outs." Firefighters don't get another chance to do

Figure 6–1. Training is the most important core principle, because there are no "do-overs" on the fireground.

better at a building on fire. They must perform their tasks and tactics correctly the first time, under time-compressed, dangerous, and stressful circumstances. And to do this, firefighters must train constantly.

And for those firefighters who are unlucky enough to work for a lazy officer who never drills, I say this—you must take it upon yourself to train and educate yourself in the core principles of structural firefighting. It's not easy, but it can be done. I know, because I once worked for a lazy officer. This guy spent his time reading the paper and watching TV instead of doing what he should have been doing, which was training the members of his ladder company. So the other firefighters on the crew and I began drilling on our own, without him. We decided that this poor officer wasn't going to stop us from being a great ladder company. We spent almost all our free time taking fire, drilling on our tools, setting up raising and lowering systems with our ropes, throwing ladders at the station, working on our search techniques, and much more. We even hijacked the rig for training. We told our officer that we needed to shop for dinner, and then, on our way to the grocery store, would stop the rig at a construction site in our district and go look at how the building was being put together. Despite our lazy officer, we found a way to train ourselves. Again, it wasn't easy, but we did it. And looking back on that time now, I think all the guys on the crew learned more than we even realized. We dedicated ourselves to training, despite our do-nothing officer, and became better firefighters.

When your rig pulls up to a house fire with a heavy smoke and flames blowing out of the windows on floor 1, smoke moving out of the windows on floor 2, and the owner in the front yard tells you that their invalid parent is upstairs in their bedroom, your crew must execute flawlessly to save that life. The only way your crew can do that is if you have comprehensively trained in the other five core principles of structural firefighting.

Whenever I watch an engine or ladder company on the fireground or performing hands-on training evolutions at our training facility, it takes me less than two minutes to know if that company trains and drills on a regular basis. The companies that do not regularly train and drill are hesitant, are unsure of themselves, and tend to make simple mistakes. The companies that do train and drill habitually and consistently are sharp and quick and know exactly what they are doing. The ladder company works together to get their ladders up quickly; the engine company driver can get the rig in pump rapidly and can get the lines charged when the hose team calls for water; the hose team stretches and moves the hose efficiently outside and inside the structure; the forcible entry team knows how to use the irons for the type of door they encounter and work together with confidence to overcome the locked door.

"But where can we train?" I hear this question all the time from firefighters and officers who have grown tired of training in or around their fire station.

My answer to this question is a simple one, yet one that many fire officers overlook: get out of your station and train in your district!

By leaving your station to train, you open up a world of possibilities. You force the crew and yourself to leave the distractions of the station and get on the rig, which is important. No more phone calls, no more TV, no more newspaper . . . you get the picture.

I remember taking the crew of Engine 38 out to a multistory, garden-style apartment house in our district and laying some dry lines, just to see what our options were. We carried 200 feet of 1¾" hose attached to a wye up to floor 3 and dropped the wye down to our driver on the street to hook up to the rig, just to see if it gave us enough operating hose for all the apartments. It did. But we also pulled the 200 foot 1¾" preconnect, took it up the stairway—making sure to snake the hose up the open well—and tied it off at the top landing, which saved many feet of hoseline compared to laying it out on the stairs. We all decided that this option would be faster than using the 1¾" wye bundle. However, we never would have known this unless we actually pulled the lines and tried it. And now we know how we would fight a fire on the top floor of this building with the 200 foot preconnected line. And the crew loved it. Two of the guys even made an ice cream bet on how to reload the wye bundle correctly in the hose bed when we were re-loading the lines—which made all of us pause and think about which was the correct way to load this bundle. Another added benefit of drilling outside the station! And the ribbing we all gave the guy who lost the bet was almost better than the ice cream.

The possibilities of drilling are endless at structures in your district. With a little help from the building manager or owner, you can really drill on almost every aspect of firefighting. Here is a brief list of the possible drills you can put together at a structure: pull dry lines off the engine to see how much hose you might need; have your acting officer place the ladder truck and then have the driver raise the aerial to the roof, or to a certain window to simulate a rescue situation; walk the stairwells of a multistory building and find out which stairways have a standpipe in them; if there is more than one stairway with a standpipe, find out if these standpipes are interconnected or need separate supplies; determine how much hose you need to cover an entire floor of the building; in high rise buildings, go over elevator operations by actually putting an elevator in Phase II firefighter control by using the elevator keys; go to several different buildings in your district and look at, and discuss, the different types of building construction; go to a building being constructed in your district or a neighboring district and see how it is being put together; go to a building with a sprinkler system and discuss how to shut down and drain the system after a fire; in front of a large building, lay dry supply line to the nearest hydrant; drive to a part of your district that is short on fire hydrants and discuss how to supply an engine company fighting a structure fire at the far end of the road or drive; if you have trains running through your district, get out and look at the types of cars (passenger or freight) that are coming through your district; on this same topic, drill on railway safety procedures and communications with rail dispatching centers; go to a gas station and locate where the emergency pump shut off is located and how you would fight a car and pump fire; drive to several single-family dwellings and look at the pitch of the roofs, to decide if a roof ladder is necessary to perform vertical ventilation; set up portable ladders at different buildings in your districts—practice going to the roof and then practice throwing ladders to reach different floors for civilian rescue; practice a firefighter rescue at a building with other companies... Really, you are only limited by your imagination here. The fact is that getting out of the station provides the best opportunities to drill (fig. 6–2).

Back to the Basics

I'm a huge fan of back-to-basics engine and ladder work. Operations firefighters need to continually practice their basic skills. And I would argue that the best way to teach and practice the basic skills is by hands-on training (HOT)—actually get out there and doing those basic skills.

Figure 6–2. Training possibilities are endless and prepare firefighters for such fires as shown here, where crew members work to deploy a 2½" line to reach a large fire in the attic of a residence.

Unfortunately, some US fire departments have forgotten the importance of basic engine and ladder work and are ignoring the fact that fires occur in structures all over the US, all the time. Instead, these fire departments and districts are spending a majority of their training time and training budgets on what can be labeled as Special Operations Training—such as weapons of mass destruction (WMD) training or terrorist event training. These misguided fire departments are applying for and relying on federal grants to fund their training budgets. They or their cities' mayors or managers are then using the fire department training budget for other "wants."

I must point out here that the federal FEMA grants were never meant to replace a department's training budget, but to supplement them. These grants were originally designed to provide fire departments around the country with much-needed equipment and training after the murderous and cowardly terrorist attacks that the United States suffered on September 11, 2001—a day we lost 343 of our brothers in the FDNY, and a day the American Fire Service WILL NEVER FORGET!

Unfortunately, some cities and departments—both large and small—saw this grant money as a windfall. These departments train all their firefighters each year on these "Special Operations," or WMD training—oftentimes at the expense of basic fire training! This is ignorant at best and criminal at worst.

Special Operations training, WMD training, and Mass Casualty Incident training are all important, as the recent terrorist attacks in Paris, France, and San Bernardino, California, have pointed out. However, to focus *an entire fire department's training schedule* on Special Operations or WMD, simply because of access to federal grant monies, is just plain wrong.

Let's be truthful here. The reality is that most towns and cities around the US will never experience a WMD or a terrorist attack. However, we know that they will experience a structure fire. All firefighters need to be training on basic engine and ladder operations at structure fires. If they don't get this basic training, then they are being set up to fail or perform poorly when they do get called to a structure fire and civilian lives are on the line.

Attention all fire chiefs, assistant chiefs, deputy chiefs and any other administration chiefs who think they know what is going on in their fire department—**take note here!** If you are not spending at least several full shifts each year riding on the busiest engines or squads or rescues or aid cars in your city or fire district, then you most likely have only a partial understanding of what is truly going on in the streets, where firefighters are working hard every shift, day-in and day-out, putting their lives and health on the line for the communities they serve. I challenge you to leave what I call "the comfort of ignorance," and ride the busiest rigs on a busy weekend shift and discover the truth for yourself. Don't rely on others telling you what is happening out on the streets—**this only gives you a filtered and sanitized interpretation, not the truth!** Get out of the office and see the reality for yourself. After all, leaders should lead from the front, not the rear.

Departments must train their firefighters on the core principles of structural firefighting and the core competencies of fighting fires in buildings of all sizes. If they don't, they are setting their firefighters up for failure, injury, or worse.

Acquired Structures

Some of the best training you can do happens when you get an acquired structure. What is an acquired structure? Basically, it is a structure slated for demolition or a complete remodel.

Someone in your fire department or fire district—preferably the chief or captain of your training division—should be working closely with your city's or town's department of construction and land use (or whatever you call the entity that oversees construction permits) to see what properties are slated for demolition or remodel, and to contact the owners and ask permission to drill in these structures.

This process is not always easy. Your department must make sure all the asbestos is abated prior to drilling in the structure, and that your city's or district's insurance will cover any accident or injury that occurs in the structure during drills. However, all this work is worth the reward!

Training in real buildings gives firefighters a chance to work in and around the types of structures they will have to work in and around their entire careers. It is much more realistic than training that occurs in the sanitized and all-too-familiar concrete structures which most cities and towns have at their training facilities. I'm not saying that drilling at the training facility is "bad" or "worthless." Not at all! Excellent training takes place at these types of facilities each and every day, all around the world. All I'm saying is that training in acquired structures adds a more "realistic" element that training in an already-memorized, sterile concrete facility just can't provide.

Acquired structure training falls into two categories: non-fire training and live-fire training. For this discussion, acquired structures fall in to two basic types: residential and commercial. I understand that there are ships, grain silos, fuel storage facilities, airplanes, stadiums, and all other specialized occupancy types. However, I am not going to get into those types here. For now, I am going to focus on the fundamentals—residential and commercial structures.

Non-Fire Training in Acquired Structures

On the non-fire training side, acquired structures can be used to practice basic engine and ladder tactics and techniques. Let's first consider residential occupancies.

On the engine side, you can have your firefighters do any one of the following: simulate fires in any of the areas of the house or apartment house—a kitchen fire, a bedroom fire, a garage or carport fire, an attic or cockloft fire, a basement fire. Have them on the engine and pull up to the house, do a 360° size-up, locate the fire, identify and isolate the flow path, make a radio size-up report (on a training channel), put the rig in pump, stretch the line to the fire, cool the simulated fire from a safe location—think transitional attack—and then enter the house or apartment and extinguish the fire. Have several engines work together to establish a water supply and get a second supply, while the hose teams pull attack and backup lines to the simulated fire (we call these MCOs, or Multi Company Operations). Practice having the engine lay reverse to the hydrant. Have your firefighters practice having both 2½" and 1¾" lines operating at the same time off the engine or a manifold. Practice blind alley lays.

For the ladder companies, you can have your firefighters practice forcible entry into the various doors of the house or apartment house using the irons and the hydraulic forcible entry tools. Black out the face pieces on their SCBA's with wax paper on the inside (or a hood on the outside) and have them practice oriented search techniques. Black out the structure with black plastic visqueen over the windows and doors, hide live "victims" (other firefighters) in the various rooms and hallways, and have teams search with the TIC and rescue the staged victims. Have your firefighters place ladders to the various rooms of the house or apartment house and practice rescuing victims. Have them throw ladders or use the aerial ladder to the roof and have them open both offensive heat holes and defensive trench cuts with the power saws. Practice deploying and using roof ladders on differing roof pitches and cutting holes from the safety of the roof ladder.

Commercial occupancies are considerably different from residential occupancies, particularly when it comes to ways in and out of the building and compartmentalization. Commercials usually have limited pathways in and out of the structure (mostly for security reasons). They often have open floor plans and are not compartmentalized like residential occupancies. Picture any big box store, and you will understand what I'm talking about.

Acquired commercial structures are just as valuable for your firefighters to train in as residential structures. On the engine side, have your firefighters pull 2½" lines, which are the preferred handlines for commercial buildings, and deploy them deep inside the structure. Have several engine companies team-up to move this line efficiently through the structure, up stairways, back into storerooms, etc. Place a manifold or tri-gated wye to the front of the building, and have the engine lay out to a hydrant supply. Deploy both 2½" lines and 1¾" lines off the manifold, flow water, and have a hydraulics drill at the pump panel afterwards. Have firefighters deploy a handline to a simulated basement fire, where a storeroom might be. Practice transitional attacks, using 2½" lines from the exterior—on the sidewalk—and then transition with a 1¾" line to the interior to finish off the simulated fire. Extend the 2½" line with a 1¾" line. Set up for a deck gun operation in front of the building, on a supply, and then transition to handlines.

On the ladder company side, black-out the structure with visqueen—or cover your firefighters' SCBA facepieces—and have your companies do an oriented search of the structure. Search off a hoseline. Use the TIC for a large area search. Ladder the roof of the building and have several ladder companies work together to cut both offensive heat holes and defensive trench cuts. Practice forcible entry through the front and back doors using all the tools at your disposal: the irons, hydraulic tools, K tool, rescue saw, axe, sledge, etc. Practice forcing roll-up doors. Smoke up the structure with a smoke machine and practice PPV with the fans. Work on firefighter Maydays and firefighter self-rescue. Set up and go through some repetitions of RIT team rescues of a downed firefighter. Practice rescue drags and below-grade firefighter rescues. Work on forcing the bars off windows and doors. Use the bolt cutters on padlocks, and figure out how to get through, around, or over chain link fencing. Ladder the windows of multistory commercial occupancies. Talk about rig placement in relation to collapse of the building. Drill on abandoning the building and orderly withdrawal. Practice rescuing firefighters from basements and below-grade parking structures.

Again, these are just a few suggestions. There are many more possibilities, limited only by your imagination.

Back when I was an officer of a ladder company, I made it a rule that my crew would throw ground ladders and use the aerial ladder every Saturday shift we worked—I called it "Ladder Saturday." On Saturdays, we would either drill behind Station 25 on the drill tower, drive to an acquired structure that I had secured for training purposes, go to an occupied apartment building in our district from whose owner I had previously received permission to conduct ladder drills. I would set up drills where I would ask the crews to rescue mannequins, or other crew members, from windows. I also asked them to shift and move all the ladders and practice deploying them on sloped ground and in tight spaces, such as alleyways and between buildings.

As you might imagine, after doing this each Saturday shift for years, my crew became experts at throwing and moving ladders. The crew also learned to work together, as a team. Over the years I watched their confidence and proficiency grow—they knew exactly what ladders to pull and how to deploy them in any given situation. And all this training paid off on the fireground, when seconds counted between life and death.

I particularly remember a fire we had one early summer morning. Engine 25 and Ladder 10 were toned out to a report of a fire on the third floor of an apartment building. The sun was just rising, and I remember having my window open, since it was already warm out.

As we were making our way west down Pine Street, dispatch came over the radio and said, "All units be advised, we have multiple calls on this fire, and there are reports of trapped victims at the windows and in their apartments."

"Sounds like we're going to be busy," Maier said over the headset from the tillerman's seat at the back of Ladder 10. "Look at the smoke."

I looked up and saw black smoke rising above the skyline several blocks away.

I knew that Woody, the veteran driver of Engine 25, would leave the front of the building for Ladder 10 and get on a hydrant. I also knew that Captain Rick Weiler, who was working overtime on Engine 25, would have the crew pull a line to the fire and get water on it as quickly as possible.

Collin, my driver, slowed the rig down as we turned onto the block where the fire was, just as he had done hundreds of times before during training and on previous fires.

"Damn," said Andy, who was riding #3 that shift. "We've got power lines near the buildings on this block... no way we can use the aerial ladder, boss" (fig. 6–3).

Before I could see the fire building, I heard the screaming through my open window—desperate screams for help that I'll never forget. They echoed through the early morning and only grew louder the closer we got.

Figure 6–3. Power lines prevent the use of using an aerial ladder.

I watched E25 leave us the front of the three-story building and move across the street to a hydrant. I also saw three women leaning out of a third story window, heavy black smoke pushing forcefully behind them. The window next to them was fully involved, flames blowing out of it and exposing the soffit of flat roof above—a tell-tale sign that we probably were going to have fire in the cockloft.

All three of the women at the window were screaming and waving their arms frantically. There was also a woman moaning on the ground below the window—she had jumped from the same window and was lying in the bushes with a severely broken leg. The three women at the window looked like they were ready to jump at any second.

Collin placed the rig perfectly, so that the back of the apparatus where the ground ladders were stored was just in line with where the victims were. This way, it would take little time to pull the ladder, throw it, extend it, and make the rescues.

"OK boys," I said, "Collin and Maierzy, throw the 35 and get those victims from floor 3. Be careful and don't let them jump on you or on the ladder before you are ready. And don't worry about the jumper in the bushes—she's alive and out. Andy and I are going inside to see if anyone's in the hallway. Got it?"

"We got it, Boss! The 35 foot ladder to floor 3 for rescue," Maierzy said. And he and Collin got to work.

I gave a report to Capt. Weiler, who had taken command, and began masking up. Just then, two men came running out of the building, their faces covered in black soot from the fire. One of these men's T-shirt was smoking, and the skin on his arms was already blistering. He said that he was burnt pretty bad. I told them both to wait across the streets for the Medic unit, and reported that there were three patients outside: the jumper and the two men.

As I was donning my facepiece at the front door of the apartment building, I looked over and saw that Collin and Maierzy had thrown and extended the 35 and placed it just below the sill, just like they had done in training countless times before during Ladder Saturdays. The women were beginning to climb down the ladder, with Maierzy up on the ladder to give them a helping hand. The bedroom where the three women were trapped flashed over just as Maierzy was bringing down the last victim.

By this time, Engine 25 team B had dropped the manifold, connected a 1¾" line, taken this handline to the third floor, and were getting ready to put water on the fire, which was now out in the common hallway. The door to the fire apartment was left open, which allowed the fire to get out into the hallway.

I checked in with Engine 25, who had just begun to apply water to the overhead of the hallway. I crawled my way down the smoke-filled hallway, and Andy and I forced the door to the apartment next to the fire apartment. I closed the door behind us. The room had smoke in it, and I used the TIC to quickly look for signs of life. We made a quick search and found no one there, so we left that apartment, marked the door with some tape so others knew we had searched it, and moved down the hallway to the next apartment.

After searching several more apartments on floor 3, the smoke began to lift, due to Ladder 4 starting up the fans and coordinating PPV with Engine 25 Team B, who now had water on the fire. Andy and I searched several other apartments, finding two occupants who were sheltering in place. We assisted them to the far stairwell, and then let Command know that we were halfway through our air supply and that we needed another company to relieve us soon and finish the primary search of the fire floor.

After Collin and Maierzy finished rescuing the victims from floor 3, they reported that there was smoke pushing out of the attic vents—there was active fire in the cockloft above floor 3. They gathered the chainsaws and roof hooks, masked up, and climbed the 35 to the roof to ventilate the fire in the cockloft. After Andy and I changed our air bottles at Ladder 10, we joined them on the roof and helped open it up. Fire blew out the heat holes we cut, and Chief Nakamichi—known to all as "Nak"—was up there with us.

"Command from Ventilation Group," Nak called over the radio, "Ladder 10 has the roof opened up, and Engine 5 is making headway on the fire in the cockloft."

The crew from Engine 5, a top-notch downtown company, was up on floor 3, pulling ceilings and extinguishing the fire in the cockloft that we had now vented. Soon nothing but steam and water was coming out of the vent holes we had made.

"Command from Engine 5, we have the fire under control," I heard over the radio.

After the fire was over, Nak came up to me as the guys were getting our equipment back on the rig.

"You and the boys on Ladder 10 did some great work this morning," Nak said. "Ladder rescues, searching for victims, cutting holes for the cockloft fire . . . You guys made it look easy, and I know it's not."

"Well Nak, my guys make it look easy because they work at it . . . a lot," I said.

"I know," he smiled. "And it shows at fires like this, where lives are on the line."

He slapped me on the back as he turned to leave. "Thanks again for all the hard work. I'll catch up with the guys back at the station and thank them personally," Nak said. "Let me know when you're ready to go back in service."

"Will do, boss," I said.

I will always be proud of what the crew of Ladder 10, C-Shift, did at that apartment fire that early summer morning. They drilled with me every shift—even on those cold, rainy winter days when I knew they didn't want to—and became one of those extraordinary crews who made everything they did on the fireground look easy.

LIVE FIRE TRAINING IN ACQUIRED STRUCTURES

I believe that live fire training in acquired structures is some of the best hands-on training there is for structural firefighters. Why? Because with live fire training in acquired structure, firefighters are fighting fire in the same type of buildings that they will be working in throughout their careers—mostly wood-frame or ordinary (brick) construction buildings, both residential and commercial. These acquired structures are much different from the typical training facility burn building—which is made of cinderblock or concrete, or a combination of both. And the fires that are set in acquired structures differ greatly from the natural gas-generated fires used in most training facilities. Wood pallets are the fuel of choice in most live burns in acquired structures, and these fires act much differently than gas-generated fires.

Now some of you may be saying that the American Fire Service has a spotty record when it comes to having training fires in acquired structures—instructors, firefighters, and probationary firefighters have died at these training fires in the past. And I would not disagree with you. But we have an NFPA Standard to guide us in how to prepare a structure and operate safely at live fire training. Any type of fire training in acquired structures must follow *NFPA 1403, Standard on Live Fire Training Evolutions*. The 1403 Standard has been written in firefighters' blood—meaning that it is a reactive Standard, and that firefighters died or were severely injured before the various chapters were added to the NFPA 1403 Standard.

Following the NFPA 1403 Standard on Live Fire Training Evolutions is not difficult, and this standard lays out exactly what needs to happen before and during live fire training.

In Seattle, our probationary firefighters end their months of training at our fire academy with a solid week of live fire evolutions in acquired single family dwellings—house fires. Our training division, where I served for a time, is very good at following and adhering to the 1403 Standard. By following the 1403 Standard, our probationary firefighters get a great introduction to how fire behaves in a wood-framed structure and how to put these types of fires out in a relatively safe environment under controlled and monitored fire conditions. By the time these probies walk out the door of our academy, they have physically worked in and around twenty or so house fires, which gives them some fantastic real-world fire experience. Sure, the fires are controlled—by limited fuel amounts, no plastics or synthetics or flammable liquids, and lower heat release rates and BTUs than exist in "real" occupied houses—but this is so the student firefighters can learn in relative safety.

There is one catch, however. Live fire training in acquired structure costs money. Why? Because NFPA 1403 dictates that before you can begin training you must provide the following: instructor training, asbestos and hazard abatement in the acquired structure, site preparation, a certain student/instructor ratio, a designated safety officer, a fire control team (staffing the safety hoselines), a Rapid Intervention Crew (RIC), Emergency Medical Services (EMS) personnel with a transport rig present, and other critical positions. But I would argue that the money spent on live fire evolutions in acquired structures more than pays for itself in terms of the experience you give the student firefighters going through this invaluable training. Simply put, there just isn't any better structure fire training than live fire training in acquired buildings.

Around the United States, which is home to some of the largest and best fire departments on the planet, live fire training has taken place in single family dwellings, duplexes, apartment houses, high rise residential buildings, high rise office buildings, and in all types of commercial structures.

Take advantage of working with your city's or district's department of construction and land use (or whatever you call the entity that oversees construction permits) and try to get the owners to allow you to conduct live fire training in structures that are going to be demolished.

A Note to All Operations Officers and Aspiring Officers about Training

As an officer, it is up to you to make sure that you are training your crew every shift. Aside from department mandated training, you—the officer in charge of your crew—need to step up and make sure that your crew is properly and thoroughly trained in all facets of the job, from fires to medical calls and everything in between. I'm talking about realistic, scenario-based, hands-on training (HOT). And no . . . realistic, hands-on training is *not* reading how to perform a hose evolution out of a book or watching another crew do it in a video.

The best part of hands-on training is that it allows you to see where the gaps are in your crews' knowledge, skills, and abilities. I can't tell you how many times that I have conducted a drill only to find out that my crew needed to work on this or that aspect of the particular task we were drilling on that shift. Training allows you to see these gaps for yourself. Once you find these gaps, then you can begin the hard work of filling them in with further training and practice.

I remember Mo, the driver of Ladder 5 B-Shift, coming to me one morning after he inspected the rig and all the equipment, just like he did at the beginning of every shift. He brought one of the chainsaws into my office and pointed out that someone on the preceding shift had accidentally put the chainsaw chain on backwards after a fire they had the night before—in other words, the chainsaw would not be able to cut and was out of service. It would have caused a huge problem for us up on a roof at an actual fire, and I was grateful that Mo did such a good job of making sure all the equipment on Ladder 5 was where it should be and working correctly. After some investigation, I found out that the probationary firefighter on the other shift had made the mistake. So I called his officer at home, explained the problem, and we discussed the solution: hands-on training for the probie and the rest of the crew. The next shift, the officer of that crew held an in-depth drill on how to break down, clean, and reassemble a chainsaw correctly. And the officer conducted this drill every month, for years, just so everyone, particularly the probie, could put a chainsaw together correctly, every time.

This brings me to the issue of skills retention and skill degradation in the fire service. The fact is that every firefighter's skills degrade over time if they are not being used *all the time*. If we are not constantly drilling on every aspect of our demanding job, then we begin to "forget" certain skills that we have not used for a while. Firefighters need the sets and reps that only constant, hands-on training can provide to combat skills degradation. Officers owe it to their crews to provide them with the opportunity to get these sets and reps, particularly on the fire side, since we just don't go to that many fires anymore. Fire officers must make sure that their crews are constantly pulling and stretching hose, throwing ground ladders, taking a hydrant, hooking up to a standpipe in a stairwell, driving the engine or ladder truck, pumping, placing the aerial on the roofs of different sized buildings, moving charged hoselines up stairwells, searching with a TIC, searching without a TIC, using fans for PPV, cutting holes in the roofs of acquired structures with saws, using the extrication tools on smashed-up cars, etc.

This problem of skill retention and skill degradation also happens on the EMS side, if your department or district is providing this service to the public, as ours does in Seattle. This means that in addition to HOT fire training, officers must be drilling their crews on all parts of basic life support (BLS) and first aid. Crews need to be proficient at AED protocols and CPR, rapid trauma surveys, splinting, C-Spine immobilization, moving patients, packaging patients for trauma, recognizing chronic medical conditions like CHF, diabetes, COPD, drug and alcohol addiction, dementia, and much, much more.

One Last Word on Training

Today, the Fire Service is being asked to do more than it has ever done in the past. The fire service is now an *all hazards service*, responding to all kinds of emergencies and problems, from structure fires to vehicle accidents to building collapses to natural disasters to pandemics to EMS calls. And the job is only getting more complex. This is why realistic, hands-on training is so important and necessary. How can we expect our firefighters to do all that they are required to do if we are not providing them with the opportunities to practice all the skills that everyone expects them to know? Effective, realistic training is the only solution.

A word of caution here. Realistic, hands-on, scenario-based training is not meant to be an experimental "testing ground" to try "new" ideas or the latest theories dreamed up by those sitting behind desks.

Just as a professional football team practices their plays during a scrimmage before an upcoming game, engine and truck companies must practice evolutions during realistic hands-on, scenario-based training. Every firefighter and fire officer needs to know the basic "play," and execute their part in that play correctly, over-and-over again, until it becomes second nature. This type of training is meant to give your crews the sets and reps of what you expect them to do on the fireground or emergency scene. But you should also build in some "what ifs" to your hands-on training, which allows your companies to practice scenarios when things don't go as planned, when things go off script. Give your officers the flexibility to change the play. Hands-on training instills confidence in the crew—they practice and get better at the skills they will perform on the fireground.

Remember that department-wide, multi-company hands-on training *is not* a time to "experiment," or "see what works." Experimentation should take place in a controlled environment, preferably at the training division or back at the station. New tactics, new ways of doing things, and new tools all need to be vetted and reviewed and evaluated and beta-tested extensively before they are ever adopted by the entire fire department.

Don't confuse your crews by experimenting during multi-company operations. Let them call and practice the plays from the playbook during HOT evolutions. Keep it simple—let your crews perform the skills they know and understand. Let them pick and run the plays from the playbook.

Operational Readiness

In this chapter, I want to discuss what it takes to be ready to work in operations—out on the streets for a fire department or fire district, either for a paid or volunteer department. Some call this "Combat Ready," or "Ops Ready," or "Duty Ready," but it is really all the same thing—being ready to do the job of being a structural firefighter!

Training is the foundation of all the other core principles of firefighting (fig. 7–1). And this is where we must start. Training is the key to being ready to respond and make a difference at:

- All types of fires: fires in residential and commercial structures, ships, airplanes, trains, all transportation vehicles, subways, grain silos, fuel storage facilities, tunnels, fires at the wildland/urban interface, flammable gas and flammable liquid and flammable solid fires, and more
- All types of emergency medical emergencies
- Vehicle accidents
- Building collapses, both big and small
- Natural gas leaks and explosions
- Electrical emergencies and fires—panels, wiring, motors, electrical vaults, downed wires
- Natural disasters: earthquakes, floods, tornados, hurricanes, and storms
- Pandemics, such as COVID-19
- Mass casualty incidents (MCIs), and Scenes of Violence (SOVs) such as shootings and stabbings
- Hazardous materials responses
- Bomb explosions, WMDs, and other terrorist events

Figure 7–1. Training mentally and physically is important for preparation on the fireground.

Fire departments need to train their firefighters to be ready to respond to all these types of incidents, for the fire service is now an **all hazards service**. (See Chapter 6.)

Some common sense is involved in mapping out a realistic training program for operations/combat personnel. The basics of structure fire training—particularly the core principles of firefighting—should never be supplanted by some other, much less-likely training, like terrorist attacks and weapons of mass destruction.

Aside from training, firefighters need to be physically, emotionally, and mentally ready to respond to all the hazards that the public we serve expects us to deal with on a daily basis out in Operations.

Being Physically Ready

Fighting fires in structures is a physically demanding job. And because of the physical demands, firefighters owe it to themselves, their crews, their families, and the citizens they serve to stay physically fit throughout their careers. And this means working out and eating a healthy diet.

The number one killer of firefighters is heart attacks. When firefighters are pushing themselves at a fire but are out of shape physically, heart attacks happen, both at the fire event, and afterwards, back at the station or at home.

It is important that our people be physically fit so that they can perform when they need to perform, on the emergency scene. The wellness and fitness programs that many fire departments have implemented, oftentimes with the help of the International Association of Fire Fighters (IAFF) and the local union, are fantastic, and should be adopted by every department, everywhere. But it is also up to individual firefighters to take responsibility for their own health and well-being. Firefighters need to get themselves into a physical training regimen and stick with it throughout their careers. If you are not working out, both aerobically and with weights, I would argue that you are not keeping yourself physically ready for the rigors of structural firefighting.

What if your fire department or fire district does not provide a wellness and fitness program?

There are many resources on the web that can point you to a physical fitness program specifically designed for firefighters. Firefighters from around the US and the world, along with the Local Unions (as part of the IAFF's Fire Service Joint Labor Management Wellness/Fitness Initiative), have designed fitness programs and have placed them online as a resource. There are websites, blogs, and videos of firefighter-specific workouts and training programs just a mouse click away.

Physical fitness not only keeps the body in shape, but it is a great way to relieve the stresses that come with the job of being a firefighter. Firefighters, whether they believe it or not, experience extreme amounts of stress—from various sources like being woken up from a sound sleep by a bell or tone at the station, or the effects of post-traumatic stress disorder (PTSD) from the tragedies we are forced to see and deal with in the course of doing the job. More on this in a moment.

Being physically ready for the job also means not coming to work with an injury or illness.

Some firefighters ignore injuries they have sustained on or off the job, and come to work when they should be a doctor's office or seeking the advice of a physical therapist. Or they come to work sick, when they should be at home in bed or being seen by their doctor.

I remember hearing how a much-loved, older lieutenant came to work one A-shift at Station 2, a busy double house in the heart of downtown Seattle. This lieutenant looked pale and tired, and said that he "didn't feel 100%" that morning. Several crew members, who were concerned about the health of their lieutenant, called the battalion chief and then called for a medic response. It turned out that the lieutenant had a major blockage, and he went in that morning for open heart surgery.

A-Shift's heads-up concern saved the life of their lieutenant, who has since retired and is enjoying every minute of his life, a life that most likely would have been cut short otherwise.

Don't hide an injury or come to work when you know you can't give your all physically—this only sets you up for further injury, failure, or worse. If you come to work sick or injured, you are taking a chance that you will not be able to perform at peak levels at a fire or vehicle extrication or some other physically demanding emergency scene.

Don't cheat yourself, your family, your loved ones, and your crew. Stay fit and stay healthy throughout your career!

Being Emotionally Ready

Firefighters need to show up for a shift emotionally ready. This means that they are not thinking about problems they may have with their family, friends, or co-workers. Family issues—such as divorce or an extremely sick child or a parent with Alzheimer's—can cause firefighters to not be present in the moment emotionally. So can money problems and other personal issues.

Being emotionally ready means that firefighters are able to put their personal problems aside and come to work ready to work and give 100%. However, life happens, and sometimes firefighters are not able to cope, or not cope very well. This is when an officer or crew member needs to step in and let the person with the problem know that they should not be at work.

I remember when I was working as a firefighter on Ladder 5, in the north end of Seattle. One of our crew members, and a good friend of mine, had arranged to take a few hours off during our shift so that he could go with his pregnant wife for an ultrasound. He was excited at becoming a father, and had another firefighter come in for him for a few hours.

Several hours later, our father-to-be came back to the station. He had a shocked look on his face. I remember asking him what was up, and he said that the doctor told them that they were having triplets! Captain Walsh, our officer, immediately called the battalion chief and sent the firefighter home for the rest of the shift.

"You need to be at home with your wife," Capt. Walsh told our father-to-be, "not here at the station." And Capt. Walsh was right. Our father-to-be did not have his thoughts on work, but on the fact that his home life was going to change drastically in the next six months or so. Capt. Walsh understood that this firefighter was not emotionally ready to perform at work. He would have been useless on the emergency scene—understandably—and could have been a liability to the crew or himself.

I can remember another time a firefighter was not emotionally ready. I was a lieutenant on Engine 18, a busy company in the Ballard neighborhood of Seattle. My driver had recently been through a long and brutally contentious divorce.

Money was very tight, and my driver was living in his truck to make ends meet. One afternoon he received a call at the station from his lawyer, who informed him that his wife was taking him back to court to garner more of his wages.

After he slammed down the phone, my driver slumped over in his chair, holding his head in his hands. "I don't have anything left to give," he kept repeating, oblivious to me and several other members of the crew who were there in the room with him. "I don't have anything left to give. Doesn't she see that?"

I left the room, called the battalion chief, and immediately laid my driver off—his thoughts were not on the job, but on his financial problems. After that, I took my driver into my office and had him make a call to our city's Employee Assistance Program (EAP), which is there to help employees going through tough times.

Luckily, one of the counselors had an appointment open later that day. My driver went down that afternoon and began seeing this counselor regularly. Talking through his problems was a real help to him, and soon he was back to work, and in a much better place emotionally. He later told me that talking with someone about his problems was the best thing he ever did.

Firefighters are human, and sometimes life can be overwhelming. Recognize this, and realize that we, too, need help sometimes. The job of being a firefighter is one of the most stressful jobs there is because we see death and traumatic injures all the time. We see things most people only see in the movies or read about in books. But for us, the images and situations are real. I believe that PTSD is a huge problem in the fire service, one that needs to be recognized for the epidemic it is. The fact is that many of us suffer from it. The American Fire Service is in the beginning stages of understanding PTSD, diagnosing it, and helping our firefighters through it. The US Military has done an admirable job recognizing that our military personnel coming back from the wars in Iraq and Afghanistan are suffering from PTSD, and they are providing help to their people. The American Fire Service needs to step up and do the same. Firefighters witness extreme pain, traumatic injuries, shootings, suffering, death, and everything in between. Firefighters deserve the same help and support that the US Military gives their people. Our people are worth it. PTSD in the fire service needs to recognized and dealt with now, not later!

Add the extreme job-related stressors of being in the fire service to everyday life stressors, and you have the very real possibility of our people experiencing emotional distress while on shift. This emotional distress manifests itself in many ways. Unfortunately, sometimes this distress causes our members to turn to drugs or alcohol for relief (more on this in a moment). And sometimes, it causes us to take our own lives.

Suicide in the fire service is a big problem, a problem that we have not talked about, but one we must face head-on. Critical Incident Stress Management Teams and suicide prevention hotlines specifically for firefighters are great first steps. But the fire service needs to do more. We need to be trained (again, training is the foundation of everything we do) on how to recognize the signs and symptoms of emotional distress and how to intervene. Officers and crew members need to be given the tools to deal with stress and emotional distress—our mental and physical well-being depends on it. We need to change our culture so that our members who seek help will not be ostracized for being "weak," or "fragile." I believe that this culture change is gaining momentum. But it is going to take an all-out effort on everyone's part. Hopefully, someday, firefighters everywhere will understand that asking for help for emotional distress is OK, and that talking to a counselor or a peer group is nothing out of the ordinary.

Being Mentally Ready

Firefighters need to be able to make good critical decisions quickly. This means being mentally ready for the job. We need our wits and smarts and common sense to respond to the problems—structure fires, wildfires, car fires, EMS calls, vehicle extrications, electrical emergencies, elevator emergencies, building collapses, storms, floods, earthquakes, hurricanes, etc.—that the public expects us to solve on a daily basis. Because, really, we are professional problem solvers.

Our minds need to be 100% in the game when we come on shift, since our lives, the lives of our crew members, and the lives of the public we serve may well depend upon it.

Being mentally ready means being prepared to handle the types of hazards we might respond to. And this goes back to training, our foundational core principle of structural firefighting. **Training** is the key to being prepared and being mentally ready to respond. We must know what to do to mitigate a hazard, whatever it may be. The only way to know how to mitigate the hazard is to have trained on it. If we know and believe that we are trained on how to handle the hazard, then we are mentally ready, and will be able to help resolve the problem we have responded to. Simply put, training gives us the confidence to solve the problems we will be called on to mitigate.

But the inverse is true as well. If we have not been trained on how to handle a hazard, then we don't have any way to know exactly what we should do to solve the problem. This leaves firefighters to make something up on the spot, which oftentimes turns out badly. Firefighters are going to act to solve a problem, even if they don't have any experience in it, because that is what they do—they solve problems, no matter what these problems are. Unfortunately, if they don't have the training, then they might become part of the problem.

I remember reading a case study of a fire department that responded to a man down in a septic tank. A firefighter went into this confined space without a SCBA to get the collapsed man, and he too succumbed to the hydrogen sulfide gas in the tank. Another firefighter went in when he saw his crew member go down, and he also died because he didn't have a SCBA on. This fire department had not trained in confined spaces, and their firefighters didn't have any experience with these types of incidents. The firefighters went in to solve a problem they had no idea how to solve, and they acted anyway. But in acting, they became part of the problem, not the solution.

Being mentally ready is having the training and the confidence to know how to solve the problem—which means training. But it also means knowing when you don't know something, and seeking expert advice in those cases—in other words, knowing not to rush in and do something just for the sake of doing something.

But what else besides training can affect our mental readiness? One negative impact on our mental readiness is the problem of substance addiction in the fire service. Alcohol and drugs—prescription, legal, and illegal—are a serious problem in the fire service. This relates back to PTSD and the extreme job-related stressors that firefighters are faced with today. Since the fire service is being asked to do so much more with staffing levels that are continually being reduced, and since we have morphed into all hazards service, it makes sense that our ranks are experiencing the severe and unrelenting stresses that go along with the reality of ever-increasing run volumes and not enough people and resources to effectively solve the problems we are expected to solve. And what happens when our firefighters are not supported with ways to cope with these severe stressors? A portion of our people begin to self-medicate by abusing alcohol and drugs. It makes sense, doesn't it? When people become so stressed, they reach for whatever they can to reduce their stress and find some comfort and relief. Some of us work out in the gym to relieve stress—this is how I deal with the stresses of the job. Some of us seek counseling from the employee assistance program (if your city provides one) or seek it on our own. Some of us talk it out with our spouses, our significant others, our friends and family, or our religious ministers or pastors. And some of us find relief in alcohol or drugs or some other destructive behavior. A firefighter who shows up to work under the influence of alcohol or drugs is not ready physically or mentally. Their judgment and reactions are impaired. And in the course of doing our job, they can be a danger to themselves, to their crew, and to the public they are supposed to serve. In short, they are a problem waiting to happen.

Fire departments need to deal with firefighters who show up to work under the influence, since they are not mentally ready to perform the job. I would argue that fire departments need to *help* these firefighters by getting them into recovery, counseling, and rehab services, and not punish or discipline them by suspending them or firing them outright for a first-time offense. The goal should be help and rehabilitation, not punishment.

The fire service must recognize that today, more than ever before, our firefighters are under extreme stress, and they need help and ways to cope with these never-ending stressors.

Investing in Your Firefighters

"The soldier is the Army," said American General George S. Patton Jr., arguably one of the greatest battlefield commanders of all time. "No Army is better than its soldiers."

This truth can be applied to every fire department around the world, big or small. No fire department is better than its firefighters.

This idea is so simple, so basic, that it often gets overlooked and forgotten by our leaders. **All fire chiefs, mayors, city managers, fire commissioners, and administrative fire chiefs take note:** Fire departments and fire districts must train, prepare, support, and invest in their firefighters to be the best they can be. Chief (ret.) John Salka of the FDNY states this simply in his excellent book on leadership, *First In, Last Out.* Salka explains that one of the foundations of great leadership is "treating your people as assets" (fig. 8–1).

Problems arise when leadership does not want to spend the time and/or the money to invest in its firefighters—when leadership does not view them as assets.

Let's face it, investing in your people takes hard work and commitment and dedication. It means coming up with standards and expectations and training. It means developing health and fitness programs, mentoring groups, and support groups for firefighters suffering from PTSD and substance abuse. It means having a long-range plan for the fire department that both labor and management craft and agree upon. It means holding your people accountable for their actions and prescribing corrective action if necessary. It means supporting and fostering positive attitudes and pride and Esprit de Corps. It means maintaining tools and equipment and apparatus so that firefighters can safely arrive at the emergency scene and have their tools and equipment work dependably every time. It means having replacement schedules for apparatus and equipment and fire stations. It means having a Post Incident Analysis (PIA) committee to review how companies performed at emergencies and to suggest areas of improvement. It means developing mentoring programs and officers' academies for every rank so that new officers can learn what the fire department expects of them in their new

Figure 8–1. Investing in firefighters means teaching them and training them in the core principles of structural firefighting.

leadership roles. It means having a cancer prevention program for your firefighters. It means spending time and money and thought on realistic, scenario-based, hands-on training. It means buying into the belief that your firefighters are your most important resource—not just saying it, but showing it with actions and money and long-term planning and institutional support.

Fire department leaders are not in the fire business or EMS business or the all hazards business. No, fire department leaders are in the people business—in the business of recruiting, training, preparing, managing, and supporting the people who respond to the fire calls, the EMS calls, the motor vehicle accident calls, the hazmat calls, the natural disaster calls, the electrical emergency calls. Our firefighters, our people—the men and women who make up the fire department—are our most important assets, and they should be viewed and treated as such.

Firefighter Down Time

As I mentioned earlier, being a career or volunteer firefighter is one of the most stressful jobs there is. And because it is such a stressful job, firefighters need down time—they need time during the shift for themselves, time to decompress, time to relax, time to de-stress.

Unfortunately, some fire department leaders mistakenly believe that if a firefighter or fire company is not on a run or call, then they need to be doing some other type of work, whether it is inspecting buildings, spinning hydrants, washing windows, etc. These "leaders" believe that if you are on the clock, then you should be "working."

I'm not saying that during their shifts fire companies should only be going on calls and conducting training drills. No. What I'm saying is that firefighters need some time during the day to relax. We need to schedule time in our day *to not do anything*, so that our firefighters can have a break and do something relaxing.

This idea of constant work may have been relevant forty or fifty years ago, when the call volumes were not what they are today and when fire-based EMS was not widely practiced in the fire service—when the fire department only responded to true fire calls. However, this constant work mentality just does not fly today. Remember, today's fire service is an all hazards service, responding to all types of emergencies, all the time.

Firefighters need time to de-stress and relax, even when they are on the job. I believe that this mental "time out" is vital to our firefighters' mental and physical health. Suicide is quickly becoming a leading cause of death of active-duty firefighters, and all the stressors that firefighters are faced with need to be lessened or mitigated.

I don't know about your fire department, but I do know that in mine, many of our companies are getting hammered with almost non-stop alarms on a daily basis, day-in and day-out, every day of the year. More than a few of our rigs are averaging twenty-five or more runs in a 24-hour shift. And this needs to be addressed, both in increasing our resources to answer the ever-increasing call volume, as well as decreasing the low acuity, or non-emergent, alarms that do not need to be answered with a fire department response.

Officers must be mindful of the constant work expectation of leadership and of the call volumes for their station and crews. Officers must give their crews a break so that they can re-energize and de-stress. The bottom line is this: down time is critical to firefighters. And everyone, from the line firefighter to the chief of the department, must understand and buy into this fact.

The Firefighter *Is* the Fire Department

So firefighters and line fire officers, what type of investment has your fire department or fire district made in you? Does your fire department treat you as an asset? Are you getting the training, equipment, and personnel you need to do your job? Is the department investing in your wellness and fitness, both physically and mentally? Does your department have a mentoring program? Does your department or fire district show you that you are their most important asset? I sure hope so. If not, I challenge you to become agents of change and transform the culture in your fire department so that the firefighters are looked upon and treated as assets instead of just another employee.

Show me a fire department or fire district that invests in its firefighters, and I'll show you a fire department or fire district that is dedicated to serving its citizens and visitors with pride and courage and professionalism and honor.

And I will even go one step further. Investing in your firefighters means teaching them and training them in the core principles of structural firefighting. These core principles are the basics, and must be understood by every firefighter at every rank. They are the foundations of what we do as structural firefighters. We must invest in teaching and training our firefighters in these foundational principles throughout their careers, from the moment they enter this noble profession until the day they retire.

9

Teamwork Gets It Done

Teamwork is the backbone of the fire service. Teamwork gets it done at every emergency scene. Teamwork is how we start and how we end. We work as a team to solve problems and to make a difference (fig. 9–1).

Firefighters work in teams for a common goal—this is the strategy of how to have a positive impact on the problem at hand, whether it be a structure fire, a vehicle extrication, a natural gas leak, downed electrical wires, a CPR resuscitation, a building collapse, rescuing victims from flood waters, or any other emergency we are called to mitigate.

To work as a team at an emergency, firefighters need to practice. Just like football players or basketball players or baseball players or soccer players

Figure 9–1. A Ladder Company works as a team on the roof of an apartment fire.

practicing and studying their opponents before a game, firefighters *must practice before the emergency* so that they know their roles and how to accomplish their tasks—which means **training**. And this brings us right back to one of the foundational core principles of structural firefighting, perhaps the most important principle: training.

Can a team ever be successful if they don't practice? They cannot! Fire departments and fire districts must provide training opportunities to their firefighters and fire officers. And as I stated in Chapter 6, we must give our firefighters realistic, scenario-based, hands-on training so that they can be have a positive impact on the emergency scene.

If you want to encourage teamwork within your crew, then you should be training. It really is this simple: training fosters teamwork.

I remember getting a call one day from the assistant chief of operations. I was a lieutenant working on Engine 18, a hard-working engine company in the Ballard neighborhood of Seattle. The assistant chief told me that he wanted me to go replace the officer on a ladder company at a busy double house in the north end of the city. He explained that the two officers—the ladder lieutenant and the engine captain—were being removed because they disliked each other and had created a toxic environment on their shift at this station. The shift was in disarray, the chief said. He also told me that he was bringing in Captain Wick to the be the engine captain.

After hearing that I would get the opportunity to work with Captain Wick, who I knew to be one of the smartest veteran fire officers in our department, I told the assistant chief that I was all-in and would be happy to help.

Soon after my discussion with the assistant chief, Captain Wick called me at home and told me that he had an idea to bring the warring engine and truck crews together.

"Steve," Captain Wick said over the phone, "we are going to drill those companies together. We are going to make them work as a team. By drilling them every shift, we will rebuild their trust in each other."

And that is just what we did. We had our companies out of the station each shift right after roll call, and we drilled and drilled and drilled—engine drills, ladder drills, pre-fires, search drills, high-rise drills, EMS drills, vehicle extrication drills, pumping and hydraulic drills, TIC assisted search drills, rope rescue drills, lifting and cribbing drills, aerial stokes drills . . . you name it, we drilled on it. All our drills were hands-on, realistic, scenario-based drills. We did not drill by sitting around a table in front of a white board or computer monitor. Instead, we were out in our district, drilling in and around buildings, at construction sites, and at vacant parking lots.

After several months of intensive drilling, Captain Wick and I noticed a change. The engine and ladder firefighters, who used to be arguing and yelling at each other, were now working hard together, joking around together, and helping each other in all aspects of their jobs, both in the station and out on calls.

I learned a valuable lesson from Captain Wick during my time at that north end fire station, that training cultivates teamwork. It was a lesson that I would never forget.

Officers must take the lead on fostering teamwork. No structure fire of any significance is ever a one firefighter or one company operation. Everyone on the fireground has a specific role to play according to your own department's standard operating guidelines (SOGs) or procedures, and firefighters working as crews and teams must work together in a coordinated fashion to bring the fire under control.

How to Promote Teamwork and the Team Concept

The team concept works on many levels in the fire service. Let's talk about teamwork at a structure fire for a moment.

Individual engine and ladder crews are a team, whether they are made up of two members or six. And both of these crews work in a coordinated fashion at a structure fire, each performing their assigned tasks to help each other and have a positive impact on the fire. So the individual firefighters on the crews must work as a team. The engine crews work together to deploy the hoselines where they are needed most, to secure a water supply, and to put water on the fire quickly and effectively. The ladder crews work together to provide forcible entry to the structure, search the structure, execute a ventilation plan to help the engine crews reach the fire and make the building behave, and throw ladders for rescue and access. All the engine and ladder crews assigned to the structure fire must work together, in a coordinated way, as a team. And the incident commander, along with the division supervisors, must work as a team to assure that all the crews assigned to and working the fire—the entire team—have the equipment and resources they need to be successful.

Teamwork is the key to success in the fire service. Foster teamwork and you will foster success at every emergency scene.

Trust

Training and trust are the pillars of teamwork.

I've spent quite a bit of time talking about training (see above and Chapter 6). Now I would like to concentrate on trust, and how it relates to teamwork.

At the individual engine or ladder crew level, the crew members must trust each other to work as a team—they have to trust that each member understands their specific job or task, and that they can perform this task efficiently and safely. The officer and firefighters need to trust one another. Without this trust, crew members will be second-guessing each other at best, or will lose confidence in themselves or their officer at worst.

Building trust must start with the officer. The officer is the leader of the crew, or team, and must be the person who begins the process of developing trust within the crew.

But how does an officer build trust within his or her crew? How can one cultivate and encourage trust to build an effective team?

Trust begins with honesty. An officer must always be honest, and speak the truth—always!

I vividly remember an engine company officer who did not speak the truth during a debriefing after a fire that did not go very well. After the fire was out, the battalion chief called all the first-in companies for a quick debrief. The battalion chief wanted to find out why it took so long to get water on a ripping basement fire in a large single-family dwelling.

The officer of the first-due engine spoke up and said that his tailboard members did not follow his instructions, a statement which was false. The truth was that the officer had failed to do a proper size-up of the house, and therefore told the tailboard members to deploy the hose to the wrong location—the front door instead of the basement door on side Charlie. The officer was trying to blame his mistake on his crew.

That officer who told a lie to make himself look good violated a sacred rule, and in so doing, lost the trust of his crew forever. From that night forward, the crew rebelled and never supported that officer in any way. They rarely talked to him, did not socialize with him, and never did anything to help him make his job easier. In the end, the officer left the company in disgrace, and his reputation as being dishonest followed him everywhere he went in his career. No one trusted him.

This officer should have admitted his mistake and taken the blame for not doing a proper size-up. This simple act of admitting his mistake in front of everyone that night would have sealed his reputation as an officer who took responsibility for his actions and who could be trusted to tell the truth.

Trust is something that an officer must work on every shift. An officer's trustworthiness is constantly being gauged by the crews he or she works with. Once lost, it can almost never be regained. So, officers and aspiring officers—remember that being honest is the foundation of trust, which every company must have to work together as a team.

Being honest and fostering trust is never easy. And sometimes, an officer must let everyone know that he or she messed up and made a mistake. Admitting a mistake, particularly in front of your crew and other companies, is difficult but necessary. I know, because I have had to admit my mistakes to my crews and others more than I would like to admit. But take heart. Everyone makes mistakes. I make mistakes on emergency scenes, but I try to address them at the time and do my best to recover from the mistake I made. Always remember that old fireground adage: It's not how badly you mess up, but how well you recover.

A Few Words on Leadership

True leaders lead from the front, not the rear!

True leaders lead the team from the front. True leaders should never ask their teammates to do tasks or take risks that they themselves would never do or take.

True leaders know what is going on the streets and in the firehouses, and don't rely on "filters"—other chiefs or captains or lieutenants—to give them a watered-down, sanitized view of what is happening in their fire department. True leaders don't sit behind their desks, in the comfort of ignorance. No, true leaders go out to the front lines, talk to the firefighters doing the work, ride the rigs, run calls, and see the truth for themselves.

Again, I use the example of General George Patton. As the commander of Third Army during World War II, Patton raced across France and drove Third Army into Germany with unprecedented speed and force. Third Army captured or killed more of the enemy, and liberated or captured more cities, towns, and villages—an estimated 12,000—than any other unit in World War II. Patton's Third Army traveled further, destroyed more of the enemy, and moved faster than any army in the history of warfare.

Patton always led his men. He did not rule them. He believed that platoons, companies, battalions, brigades, divisions, and armies were all like a piece of cooked spaghetti. "You can't push it," he said, "You've got to get in front and pull it." Patton insisted that his senior officers, himself included, "get up front," and get out of headquarters. He once told his senior staff:

> *I want every member of this staff to get up front at least once every day. You will never know what is going on unless you can hear the whistle of the bullets. You must lead the men. It is easier to lead than to push.*

Patton always led from the front, and he shared in the hardships and risks of his frontline troops.

This is true leadership—leading from the front. There simply is no other way to lead if you wish your crew or battalion or division or department to be successful. If you are not leading from the front, you are following, unable to see or truly understand what is happening on the streets, on the front line, where your firefighters are working day-in and day-out. And if you are following, then you are not a leader.

Don't be a follower. Get out front and lead!

Do the Right Thing

The leader of any successful and effective team must always do the right thing. But what does this truly mean? What is "doing the right thing?"

Doing the right thing means training your firefighters every shift. It means providing training that is relevant, thought-provoking, and mentally and physically challenging. It means working and training alongside your firefighters, never asking them to do anything that you yourself would not do. It means leading by example and volunteering to be the first firefighter through a training prop. It means pitching in and helping do the housework around the fire station every shift. It means answering the doorbell at the fire station and listening to the citizen who has come with a question or a concern, no matter how trivial it may seem.

Doing the right thing means taking the time to be present in the moment and to listen—both to the firefighters on the crew, and to the public whom we serve. It means being there to help the crew reload the hose beds on the engine after a fire. It means being honest and trustworthy and diligent and on time. It means working hard and being the last person to leave the scene and get back on the rig. It means keeping yourself in top physical shape so you can be an example that others look up to. It means following and supporting your department's policies and guidelines. It means having a positive, can-do attitude. It means arriving early for your shift. It means understanding that everyone deserves a second chance after making a mistake. It means continually learning the craft and science of structural firefighting. It means passing on your knowledge to the next generation of firefighters in your department. It means being able to laugh at yourself when you make a mistake in front of others.

And lastly, doing the right thing means always doing your best at the time, no matter what.

The Simple Act of Listening

Good leaders know how to listen.

It seems like such a simple thing—to actively listen. You would be surprised to find out that many leaders don't know how to listen to the people they are supposed to be leading. Instead, they pretend to listen, and do not pay attention to what the troops are saying. Or they always talk, and never let anyone get a word in edgewise.

Great leaders actively listen to what their people are saying and respect their input. Listening—really listening—fosters trust and cooperation, and it gives your people ownership of ideas and a shared vision, which is ultimately what every leader should want to promote.

Listening to your crew or your station allows you to get new perspectives and ideas from your people—people who may have a better solution or a better way of doing something. No leader has all the right answers. Great leaders empower their people by listening to their ideas, complaints, and solutions. Great leaders actively listen.

I'm not saying you must always agree with your people. Sometimes you have to agree to disagree. But by truly listening, you liberate your people and build that trust that I spoke about earlier.

Stop talking, and listen.

Focus on Results Over Process

Today, many fire districts around the world are implementing standard operating guidelines (SOGs) or standard operating procedures (SOPs) for many of the types of runs they may respond to—SOGs or SOPs for structure fires, vehicle extrications, natural gas leaks, Haz-Mat responses, Scenes of Violence, elevator rescues, water rescues, etc. Having a general playbook that is not too restrictive and confining in terms of how to mitigate these emergencies is good—everyone knows the basic objectives and the desired results.

However, some departments are writing SOGs and SOPs that are focused too much on the Process, sacrificing desired Results for a step-by-step script. The problem with focusing on the Process rather than the desired Results is that when the emergency does not conform to the step-by-step script, or veers off the script, then the officers and firefighters may not know what to do, which could lead to disaster.

A good analogy would a football quarterback who was given an offensive play to run by his coaches on the sideline. When the quarterback walks up to the line of scrimmage, he notices that the defense is not lining up the expected way, and that the defense is positioned to wreak havoc on the offensive if they run the called play. So instead of following the process and running the play that the coaches called for, the quarterback audibles at the line, and calls a new play, one that will take advantage of the defense's current formation on the field. The offense, led by the quarterback, executes the new play, which takes the defense by surprise, and makes a forty yard gain deep into the opposing teams territory. The desired Result—an offensive gain—took precedence over the Process of the offense running the coaches' play that would have resulted in a negative outcome, such as a turnover or yardage loss.

Fire chiefs and assistant chiefs—listen closely here. When writing SOGs or SOPs, keep them Results focused! Don't fall into what I call **the process trap.** Give your firefighters and fire officers the tools and the training to know the desired Results, and give them the latitude and provide them with some options to make those Results happen on the fireground or emergency scene. Allow your officers to audible the play, so they can deal with any unexpected problems that might arise.

The SOGs or SOPs for a structure fire should be Results-driven and provide a loose framework that focuses on these goals. An excellent example of this would be following the tactical objectives of Rescue, Exposure protection, Confining the fire, Extinguishment, Overhaul—RECEO—for all structure fires. How the fire officers and firefighters achieve these objective or results should be up to them at any given structure fire. This allows them the freedom to use the tools and training they have hopefully been given to solve the problem—the structure fire—that they are facing.

The take-away is this: do not let your people be constrained by a Process or script that is unrealistic and does not take into account the dynamic and ever-changing unknowns on the fireground.

FINAL THOUGHTS

As I stated at the beginning of this book, aggressively fighting fires in and around structures has always been a difficult and dangerous job.

Firefighters often get hurt or they can die at building fires because they lack the basic knowledge or understanding of one or more of the core principles of structural firefighting. These core principles are fire behavior, building construction, strategy, tactics, safety, and training.

Fire departments must constantly train their firefighters and officers in these core principles if they want them to be safe and effective at structure fires. Remember that training is the foundation of all the other core principles, and it must be realistic, scenario-based, and hands-on (fig. 10–1).

Figure 10–1. Training is the foundation of all other core principles and must be realistic, scenario-based, and hands-on. Training ensures safety and effectiveness at structure fires.

Leadership and building an effective team or crew is never easy. It takes time, patience, and trust. It also takes a common belief in the mission, which, for the fire department, is serving the public. Leadership requires that the leader must lead from the front and be an example that his or her crew or battalion or department wish to emulate. Leaders must do the right thing, always!

This brief study is only meant to serve as a beginning, a starting point, for you, the reader. I've only scratched the surface of the six core principles of structural firefighting in the preceding chapters. You must never stop learning during your time as a structural firefighter. If you come to a point where you mistakenly believe that you know everything there is to know about fires in and around buildings—watch out—because you have just fallen into the complacency trap. Remember, complacency kills firefighters!

So, What Now?

Now it's up to you. Be the firefighter, the company officer, the battalion chief, or the deputy chief who is always seeking knowledge, always trying to discover more about the craft and science of fighting fires in buildings.

I leave you with this final challenge: Never stop learning and studying all that you can about this wonderful and dangerous and fulfilling profession you have chosen—the noble and courageous profession of structural firefighter.

RECOMMENDED READING

Angulo, Raul. Engine Company *Fireground Operations*, 4th Ed. Massachusetts: Jones & Bartlett Learning, 2020.

Axelrod, Alan. *Patton on Leadership, Strategic Lessons for Corporate Warefare.* New Jersey: Prentice Hall, 1999.

Avillo, Anthony. *Fireground Strategies*, 2nd Ed. Oklahoma: Pennwell, 2008.

Bingham, Robert C. *Street Smart Firefighting*, Virginia: Valley Press, 2005.

Brannigan, Francis L., and Glenn Corbett. *Brannigan's Building Construction for the Fire Service*, 4th Ed. Massachusetts: Jones and Bartlett, 2008.

Brennan, Thomas F. *Tom Brennan's Random Thoughts.* Oklahoma: Pennwell, 2007.

Ciampo, Michael. "On Fire: Life, Fire, and Operations." *Fire Engineering*, May 2010. And the last page of every *Fire Engineering* Magazine for the last 10 years.

Corbett, Glenn, ed. *Fire Engineering's Handbook for Firefighter I and II.* Oklahoma: Pennwell, 2009.

Dunn, Vincent. *Safety and Survival on the Fireground.* New Jersey: Fire Engineering Books and Videos, 1992.

Dunn, Vincent. *Strategy of Firefighting.* Oklahoma: Pennwell, 2007.

Fields, Aaron. *Bread & Butter Operations: Fire Streams.* Fire Engineering Books & Videos, April, 2013. DVD.

FirefighterCloseCalls.com. http://www.firefighterclosecalls.com/

Gagliano, Mike, Casey Phillips, Phillip Jose, and Steve Bernocco. *Air Management for the Fire Service.* Oklahoma: Pennwell, 2008.

Giselsson, Krister, and Mats Rosander. *The Fundamentals of Fire.* GIRO-Brand AB, 1978.

Grimwood, Paul. http://www.firetactics.com

Gustin, Bill. Any article he has written for *Fire Engineering* magazine

Marsar, Stephen. "Survivability Profiling: Are the Victims Savable?" *Fire Engineering*, Dec. 2009: 69–72

Mittendorf, John. *Truck Company Operations*, 2nd Ed. Oklahoma: Pennwell, 2010.

Mittendorf, John, and Dave Dodson, *The Art of Reading Buildings.* Oklahoma: Pennwell, 2015

The National Institute for Occupational Safety and Health (NIOSH). https://www.cdc.gov/niosh/index.htm

Norman, John. *Fire Officer's Handbook of Tactics*, 3rd Ed. Oklahoma: Pennwell, 2009.

Salka, John, and Barret Neville. *First In, Last Out.* New York: Portfolio, Penguin Group, 2004.

About the Author

Steve Bernocco is a twenty-nine-year veteran of the Seattle Fire Department, where he currently serves as captain of Engine Company 38, located near the University of Washington. He is one of the authors of *Air Management for the Fire Service*, the pioneering text on fireground air management and firefighter safety. Captain Bernocco speaks and teaches nationally and internationally about fireground strategy, tactics, and safety. He is dedicated to his family, his firefighters, and to the men and women of the Seattle Fire Department.

About the Photographer

John Odegard volunteers as a photographer for the Seattle Fire Department, which routinely utilizes his photos for everything from training to investigations to public relations. John's father served as a captain in the Tacoma (WA) Fire Department.

Ladder 10 and Aid 25, C-Shift

INDEX

E

F